Status and Trends of
GLOBAL BIODIVERSITY

Status and Trends of GLOBAL BIODIVERSITY

Avinash Tyagi

RAJAT PUBLICATIONS
New Delhi

Status and Trends of Global Biodiversity
by *Avinash Tyagi*

This edition is published by Rajat Publications
4740/23, Ansari Road Daryaganj,
New Delhi - 110002

ISBN: 978-81-7880-254-1

Edition 2026

Printed and bound in India

RAJAT
PUBLICATIONS

4740/23, Ansari Road, Daryaganj,
New Delhi- 110002 (India)
Mob:- +91-9811484973, +91-9811105389
E-mail: rajat_publication@yahoo.com

Preface

The word 'biodiversity' is a combination of two words: biological and diversity. It refers to the variety of life on Earth. Biodiversity encompasses all the living things that exist in a certain area, in the air, on land or in water: plants, animals, microorganisms, fungi. The area considered may be as small as your backyard compost heap or as big as our whole planet. All living things are connected to other living things and to their non-living environment. If one tiny species in an ecosystem becomes extinct, we may not notice, or think it's important. But the biodiversity of that ecosystem will be altered, and all the ecosystems that the species belonged to will be affected.

Biodiversity found on Earth today is the result of four billion years of evolution. It constitutes the most important working component of a natural ecosystem. There are three aspects to biodiversity: species diversity, genetic diversity and ecosystem diversity. All three interact and change over time and from place to place. Biodiversity helps maintain ecological processes, creates soils, recycles nutrients, has a moderating effect on the climate, degrades waste, controls diseases and above all, provides an index of health of an ecosystem. Providing food, medicines and a wide range of useful products, it is the natural wealth that exists on land, in freshwater and in the marine environment.

The rich biodiversity of the world, however, has been under serious threat from a variety of human induced factors. As the human population grows, and people encroach on more and more land, habitats can become fragmented and degraded. Another threat to biodiversity is the issue of climate change. There is worldwide concern over global warming. Pollution of our air and waterways also affects biodiversity. Overharvesting depletes the Earth's stocks of animal and plant resources and the ill-considered introduction of exotic species also has a negative effect on biodiversity. Although many of the threats to biodiversity are the result of human activity, it is only human can make positive changes — adopting attitudes to the Earth's rich resources that will respect and maintain this diversity.

The purpose of this book is to provide an overview of the changing status of the world's biological diversity. It provides general information on the nature and scope of biological diversity, and broad trends at genetic, species and ecosystem level. It also covers a wide range of issues such as the causes and consequences of biodiversity erosion; the strategies of biodiversity conservation; the global biodiversity convention and its impacts. Being an all-encompassing presentation on biodiversity, the book is destined to become a highly useful reference tool for academicians, researchers, scientists, administrators and ecoplanners.

Editor

Contents

1

Biodiversity and Ecosystem

Earth include about 10 million different species. The wide-ranging decline in biodiversity results largely from habitat modifications and destruction, increased rates of invasions by deliberately or accidentally introduced non-native species, over-exploitation and other human-caused impacts. On a global scale, even at the lowest estimated current extinction rate, about half of all species could be extinct within 100 years.

Such an event would be similar in magnitude to the five mass extinction events in the 3.5 billion year history of life on earth. On local and regional scales, biodiversity declines are already pronounced in many areas, especially where natural ecosystems have been converted to croplands, timber plantations, aquaculture and other managed ecosystems. The diversity of these managed ecosystems is often low, and species composition very different, compared with those of the natural systems they have replaced. The earth's living organisms contribute to human welfare in a variety of ways.

First, humans derive from them goods and products essential to life, including food, medicine, and industrial products, genetic resources for crop breeding, and natural pest control services. Such benefits can be viewed as the market values of biodiversity because they are readily tied to our economy and often can be assigned a dollar value in the marketplace.

Second, biodiversity has nonmarket values that can be expressed in terms such as knowledge, aesthetic, existence and other values. These non-market values of biodiversity are difficult to quantify, but are, for many, sufficient justification for preserving biodiversity independent of market values.

A third category of value is ecosystem services. The organisms that live, grow, reproduce, and interact within ecosystems help to mediate local and regional flows of energy and materials. Energy flow refers to the capture of light energy by green plant or algal photosynthesis and its dispersal as chemical energy throughout the food web to plant- or algal-feeding animals, predators, and eventually decomposers.

The flow of materials involves the recycling of carbon, nitrogen, phosphorus and other elements between living organisms and the air, water, and soil. These biologically mediated energy and materials flows contribute to many ecological or life support services that benefit human welfare such as greenhouse gas regulation, water treatment, erosion control, soil quality control, and plant growth. Ecosystem services can also include cultural benefits, such as religious, aesthetic, recreational, or inspirational values that humans derive from ecosystems. Determining whether biodiversity per se is important to ecosystem functioning has been difficult, partly because many of the factors such as habitat conversion that reduce local biodiversity also directly affect many ecological processes, masking the more subtle impacts of species loss on functioning.

Biodiversity

Although every organism contributes to ecosystem processes, the nature and magnitude of individual contributions vary considerably. Research in biodiversity places much emphasis on the uniqueness of individual species and their singular contributions to ecosystem services. Yet most ecosystem processes are driven by the combined biological activities of many species, and it is often not possible to determine the relative contributions of individual species to ecosystem processes. Species within groups such as grazing mammals, large predators, perennial grasses, or nitrogen-fixing microbes may therefore be functionally similar despite their uniqueness in genes, life history, and other traits.

Groups of species that perform similar roles in an ecosystem process are known as functional types or functional groups. Species may also be divided into functional types based on what they consume or by trophic status. Within trophic groups, species may be further divided according to life history, climatic or nutrient needs, physiology or other biological traits. Researchers may place a species into several different functional categories depending on the ecosystem process they are studying. Because species can vary dramatically in their contributions to ecosystem functioning, the specific composition or identity of species in a community is important.

The fact that some species matter more than others becomes especially clear in the case of 'keystone species' or 'ecosystem engineers' or organisms with high community importance values. These terms differ in usage, but all refer to species whose loss has a disproportionate impact on the community when compared to the loss of other species. For example, a species of nitrogen-fixing tree,

Myrica faya, introduced to the Hawaiian islands has had large-scale effects on nitrogen cycling, greatly increasing the amount of this essential plant nutrient in soils where the tree invades.

The nitrogenfixing lupine Lupinus arboreus also enriches soils and, as a consequence, encourages invasions of weedy grasses. Among animals, moose through their dietary preferences greatly reduce soil nitrogen levels and also influence the succession of trees in the forest. Beavers, too, through their feeding and dam-building not only alter soil fertility and forest succession but increase the diversity of ecosystems in a landscape. Even termites play critical roles in soil fertility and other ecological processes in many arid grasslands.

Ecosystem Functioning

Ecosystem functioning reflects the collective life activities of plants, animals, and microbes and the effects these activities—feeding, growing, moving, excreting waste, etc.—have on the physical and chemical conditions of their environment. A functioning ecosystem is one that exhibits biological and chemical activities characteristic for its type. A functioning forest ecosystem, for example, exhibits rates of plant production, carbon storage, and nutrient cycling that are characteristic of most forests. If the forest is converted to an agroecosystem, its functioning changes.

Ecologists abstract the essential features of an ecosystem into two compartments, the biotic and the abiotic. The biotic compartment consists of the community of species, which can be divided functionally into plant producers, the consumers that feed on producers and on each other, and the decomposers. The abiotic compartment consists of organic and inorganic nutrient pools. Energy and materials move between these two compartments, as well as into and out of the system. Ecosystem processes are quantified by measuring rates of these movements.

Ecosystem functioning, in turn, is quantified by measuring the magnitudes and dynamics of ecosystem processes. Ecosystem functioning results from interactions among and within different levels of the biota, which ecologists describe as a 'nested' hierarchy. For example, green plant production on land is the end product of interactions of individual plants nested within populations; interactions among populations nested within a single species; interactions among a variety of species nested within a group of functionally similar species; and so on up to the level of interactions between different types of ecosystems nested within landscapes.

Relationship Between Biodiversity and Ecosystem Functioning

Since Darwin, prominent biologists have hypothesized about the relationship between biodiversity and ecosystem functioning. More recently, concerns about increasing loss of biodiversity and questions about resulting degradation of ecosystem services have stimulated unprecedented observational, theoretical, and experimental studies.

Observational studies

It might seem that observational studies comparing one ecosystem type with another, or comparing similar ecosystems at different locations, could provide ready answers to questions about the impacts of species richness on ecosystem processes. But these studies have invariably proven problematic. For example, an ecosystem such as a tropical forest or a coastal wetland may vary from one site to another not only in species number and composition, but also in physical and chemical conditions such as soil type, slope, rainfall, or nutrient levels.

Comparing different ecosystems is likely to yield an unclear result because the response to variations in biodiversity cannot easily be distinguished from responses caused by variations in environmental and other factors. It is possible, though difficult, to control statistically for such potentially confounding factors.

Experimental Studies

Experimental studies, if well-designed, can minimise the confounding factors that plague observational studies. Experiments can provide insights not only into the relationships between biodiversity and ecosystem functioning but also into the possible mechanisms behind the relationships. Studies to date have ranged from large outdoor experiments and trials in large controlled environment facilities to modest-sized pot experiments and tests in small laboratory microcosms.

This research has attempted to address two different questions about the link between biodiversity and ecosystem functioning. First, how are levels of ecosystem functioning affected by changes in biodiversity, particularly species richness? Second, how are the dynamics of ecosystem functioning, particularly the resilience and stability of processes, affected by changes in biodiversity?

Results from many recent experimental studies conducted in North America and Europe demonstrate that ecosystem productivity increases with species richness. These studies range from large outdoor experiments to controlled laboratory experiments conducted in growth chambers, greenhouses, or small containers. Outdoor experiments such as those conducted in grasslands on nutrient-poor serpentine soils at Stanford, California and on prairie grasslands at Cedar Creek Natural History Area, Minnesota, work with plant communities similar to those found in nature, but researchers vary the number of plant species from one experimental plot to another. This approach is also used in the Biodepth experiments, in which seven European countries have established outdoor plots that range in plant diversity from low species numbers to the average numbers typically found at each site.

More recent laboratory experiments in Europe and North America have begun to examine the impact of other components of biodiversity, such as the diversity of soil microorganisms, on plant production and the role of bacteria, predators, and herbivores in freshwater microbial communities. All of these studies show that ecosystem functioning is decreased as the number of species in a community decreases. Declines in functioning can be particularly acute when the number of species is low, such as in most managed ecosystems including croplands or timber plantations.

In addition, recent experimental studies in grasslands indicate that the effects of biodiversity on production can depend on both the number of functional groups present and the identity of the plant species. Other studies have shown that loss of functional groups from a food web, or reductions in the number of species per trophic group can also cause declines in ecosystem functioning. Finally, another study has shown that some species of plants may be more or less productive or show no response at all to changes in the diversity of their communities, even though total community productivity is, on average, lower at lower diversity.

Studies on plants have been particularly revealing and support results from recent theoretical models which predict that decreasing plant diversity leads to lower plant productivity. These models predict that diversity and composition are approximately equal in importance as determinants of ecosystem functioning. Two possible mechanisms have been identified to explain why levels of ecosystem functioning increase with increasing biodiversity.

First is the 'sampling effect': When the pool of species available in a region contains individual species that vary in productivity and other contributions to ecosystem functioning, then species-rich ecosystems have a higher probability of containing species with high levels of functioning.

Second is the 'complementarity effect': This occurs when increasing diversity results in increasing numbers of species that are complementary rather than competitive in their use of resources, exploiting different niches, such as rooting depths, and allowing more effective use of available resources.

Impact of Biodiversity on Stability

Few experimental studies of the impact of biodiversity on stability have been attempted, largely because stability is a long term attribute of a system and testing for it requires either long-running experiments or experiments with short-lived organisms. In the one available long-term ecological field study, however, reductions in plant species richness also lowered the resistance of grassland production to drought. Predictably 'lower year-to-year fluctuations in community productivity' was also significantly lower at lower diversity.

In addition, studies of microbial communities in small experimental chambers have also shown that fluctuations in ecosystem functions such as productivity can be greater when species richness is reduced. Thus, the loss of diversity causes a loss of ecosystem stability. Several mechanisms could account for these results. One mechanism comes from the ability of competing species to replace or compensate for one another and thus minimise, at higher diversity, the ups and downs in functioning. Another mechanism is the 'portfolio effect' a theory which suggests that cumulative properties such as ecosystem functioning show less severe fluctuations in systems with many species, much the way investment portfolios of varied stocks have lower long term variance than portfolios of one or a few kinds of stocks. Three points emerge from this growing body of research.

First, declining species richness can lead to declines in overall levels of ecosystem functioning. This is especially pronounced at lower levels of diversity. This finding is particularly relevant to current ecological change, since most ecosystems are being transformed into managed systems which typically contain only a few dominant species, whereas the natural ecosystems they replaced typically contained tens to hundreds of species.

Second, at least one species per functional group is essential to ecosystem functioning. Having more than one species per functional group may or may not alter overall levels of ecosystem functioning, but it may nevertheless insure against loss of functioning in times of disturbance if species within functional groups are able to replace or compensate for one another.

Third, the nature of an ecosystem's response to declining biodiversity is dependent on community composition, that is, on which species are lost and which remain. Research to date, however, has not identified any clear rules allowing us to predict in advance the impacts of the loss of any particular species on ecosystem processes.

Although these three points have been repeatedly observed in a wide variety of experiments, there is still debate about the mechanisms behind them. Research into the link between biodiversity and ecosystem functioning is a new discipline and much work remains to be done. Research to date strongly supports the idea that ecosystem functioning is sensitive to changes in local species identities, community composition and diversity.

Although current studies are limited in scope, they do demonstrate that plant production, nutrient use, nutrient leaching, soil fertility, and the predictability and stability of ecosystem processes can falter in the face of reductions in biodiversity. Despite this progress, several areas of uncertainty remain to be investigated. What are the effects of changes in biodiversity at scales other than species or functional groups? Most studies involving biodiversity and ecosystem functioning have focused only on changes in the number and variety of species and/or functional groups. Yet many important ecological processes occur at the landscape level, and current studies strongly suggest that landscape level alterations of biodiversity affect ecosystem functioning. There is a need for experimental research that manipulates biodiversity at both larger and smaller scales.

Is current knowledge applicable to all ecosystems? Studies to date have examined primarily isolated ecosystems. Future experiments across multiple ecosystem types will be needed to test whether findings from lakes or grasslands, for example, can be applied more widely. This approach is already being tested in Biodepth, a pan-European biodiversity-ecosystem functioning experiment that may serve as a model of the kind of experiments needed. At eight field sites across Europe, Biodepth researchers have created grass-herb ecosystems with varying levels of biodiversity drawn from local species pools. Results of these studies will expand to the landscape level our understanding of the relationship between biodiversity and such ecosystem processes as production, decomposition and nutrient retention.

How important is diversity at all levels of the food web to ecosystem functioning? With the exception of some studies conducted inside laboratory growth chambers, most experiments to date have considered only plant species diversity and not variations in the numbers of herbivores, carnivores, parasites, decomposers and other players in the food web. Yet these creatures not only comprise the most numerous portion of the earth's biota but are also significant players in the flow of materials and energy. Experiments that involve multiple levels of the food web are critical to expanding our understanding of the ecological consequences of biodiversity loss.

How will other global changes interact with changing patterns in biodiversity and ecosystem functioning? Currently, few experiments are explicitly examining interactions among such factors as increased atmospheric carbon dioxide, increased untraviolet-B radiation, increased nitrogen deposition, global warming, habitat fragmentation, and changing patterns of biodiversity. Experiments considering all these factors at once are impractical. Yet one project that examines interactions of three of these factors is currently in progress at Minnesota's Cedar Creek Natural History Area.

The experiment, Biocon, manipulates plant diversity, carbon dioxide, and nitrogen in experimental grassland plots. What are the economic consequences of ecosystem responses to changing biodiversity? Currently, economic valuations have focused on market values of either ecosystem services or biodiversity. Future analyses which integrate both biodiversity and ecosystem functioning may provide a better understanding of the potential economic impacts of biodiversity loss.

2

Genetic Diversity

Genetic diversity is based on variation between genes, i.e. in the functional units of heredity in organisms. Not all the genetic material has a role in heredity, and in some organisms a large proportion of DNA is highly repetitive in sequence and has no known function. The features of an organism are determined fundamentally by the genes received from the previous generation, by the way these genes are expressed and interact, and to an extent by environmental effects on the organism.

New genetic diversity arises when chromosomes are not copied with absolute accuracy (i.e. undergo mutation) and through reassortment of genes on chromosomes when sex cells (sperm and egg cells) are being produced. In this latter process, different forms of genes at corresponding positions on maternal and paternal chromosomes can become exchanged, and entire chromosomes of maternal and paternal origin become sorted into different combinations. If the features determined by new mutations and new permutations of genetic material improve survival of individuals bearing them, the frequency of these genes in the population will increase.

Genetic diversity is important as it provides the raw material of evolution and because it enables adaptation and change in organisms. High levels of genetic diversity should allow species to be flexible in the face of environmental change, whereas low genetic diversity, for example in a small isolated population, tends to increase the risk of extinction. Genetic diversity is also critically important for the continuing ability of human societies to derive economic and social benefits from biodiversity.

The variability is an insurance policy that protects against risks that could reduce such benefits, for example widespread pest outbreaks or fluctuations in crop production from year to year. Complex biochemical techniques can be used to measure the frequency and distribution of different gene products in species' populations, or genetic diversity may be assessed at organism level, for example in terms of the measurable production or pest resistance qualities of different crop varieties.

Bacteria differ from other organisms in that they can reproduce at a very high rate by dividing in two. They also exchange genetic material, but this sexual process is not associated with reproduction as it is in other organisms; instead, two bacteria in contact may directly exchange genetic material, or a single bacterium may take in DNA deposited in the environment by another bacterium. These exchanges, the latter in particular, are not always constrained by the kind of barriers that in higher organisms restrict exchange of genetic material to members of the same species.

Genetic material can also be passed between bacteria by viral infection. The apparent frequency with which genes move between populations of bacteria means that these organisms are able to generate high levels of new genetic diversity for natural selection to act upon, and favourable genetic material can spread very quickly in rapidly reproducing populations. This explains why resistance to antimicrobial substances can arise so readily in bacteria.

Human activities readily lead to change in the genetic diversity of populations of wild or domestic organisms. In extreme cases, genetic diversity is reduced to zero when a species is rendered extinct. More usually, it is reduced to some extent when populations of species are reduced in abundance or distribution. Such loss of genetic diversity, particularly if brought about by human activity, is often termed genetic erosion. The extent of reduction depends on the amount of diversity in the declining population and the way it is distributed geographically.

Although levels of genetic diversity within such species would in some circumstances be expected to recover over time, particular genetic material that might be important for future adaptation may well be lost completely. It would be possible, although not practicable routinely, to demonstrate genetic erosion at gene level. More usually it is evident at the species level, when measurably distinct populations (perhaps some local variety of crop plant) are lost or reduced in abundance, or is simply assumed to have occurred when an area of species-rich habitat is cleared. Because of the prevailing high rate of habitat modification, it can safely be assumed that genetic erosion is a very common and widespread phenomenon; however, not every loss of a local population is an instance of genetic erosion, because that population may have held no elements or combinations of genetic diversity not found elsewhere.

Human activity can also increase genetic diversity in species populations. By a process of artificial selection, humans have been indirectly but purposefully promoting genetic change in species during a period of more than

10,000 years, resulting in the current world diversity of domesticated crops and livestock. Artificial selection involves managing the reproduction and survival of individuals within populations of useful plants and animals so as to preserve and make abundant those lineages that possess particularly useful features, such as high pest resistance or milk yield. The process is indirect in that it is focused on tangible features rather than on the genetic material from which such features originate.

The activities collectively known as genetic engineering also involve manipulation of existing genetic material, creating new gene combinations with the aim of improving key features of organisms used by humans. Genetic engineering is in one sense only an extension of traditional breeding practices, in that it relies on naturally occurring elements of diversity. However, it is fundamentally different in other important respects. Not only does it involve making direct modifications to the actual genetic material of organisms, but also genetic material from different kinds of organism usually isolated reproductively from each other, can be brought together and perpetuated in new lineages.

The more radical forms of genetic engineering have only been developed during the 1990s but already have had considerable social impact. The techniques may have great potential to improve efficiency, volume or quality in agricultural and other production processes, and these potential benefits could be of particular value to countries at risk of food insecurity. However, they also raise significant ethical and practical concerns, which have been expressed by scientists and by public opinion in both developed and developing countries.

Among the practical concerns, attention has focused on the possible effects of genetic material moving from genetically modified (GM) sources, particularly field crops, into other organisms. Regarding plants, the risk of this occurring depends partly on whether the crop is an inbreeding or outbreeding species, and whether wild relatives of the crop are grown in the area. The evidence available on possible effects at other levels in the food chain, for example on plant-feeding insects, is sparse but indicates that concern may be warranted. The use of genes conferring resistance to antibiotics as marker genes has caused concern because of the potential for increasing resistance in naturally occurring bacteria.

It is widely believed that some movement of genetic material from GM sources into other organisms is inevitable in the long term; the level of risk that can be tolerated depends on the balance of benefits and costs. These questions of responsible management of GM resources and technology have given rise to the new field of "biosafety." Human activity has also unintentionally led to increased genetic diversity, particularly among bacteria. In such cases a form of artificial selection is imposed on bacterial populations when they are exposed to antibiotics. It has been shown, for example, that routine use of antibiotics as prophylactic or growth promoting agents in intensive agriculture, i.e. not just for therapeutic purposes, has led to emergence of bacterial strains that are resistant to antibiotics, and these strains can infect humans.

3

Species Diversity

The fundamental practical importance of trends in genetic diversity, biodiversity change has mainly been assessed in terms of declining populations and species, either individually, or collectively, when manifest as loss of habitats or reduction in area of ecosystems. Historically, the impetus for much conservation activity has been the drive to prevent the decline and extinction of individual species, with significant emphasis on species that are large and charismatic. The primary benefit of this approach may be that large organisms, terrestrial vertebrates in particular, generally require large areas of suitable habitat, and if such areas can be managed to minimise risk, other species in the system may be safeguarded.

There has always been special concern about extinction because of its irreversibility, and the loss of a species will entail loss of unique elements or combinations of diversity at gene and organism level. In this regard, the fossil record demonstrates two important facts. Firstly, that although relative rates have varied greatly, over geological time as a whole there has been a net excess of species originations over species extinctions (i.e. biodiversity has increased). Secondly, that virtually all the species that have ever existed are now extinct, and the extinction of every species is a natural and expected event.

Self-evidently there must always have been species at risk of extinction, i.e. "threatened species." It is difficult for many reasons to keep track of species extinction in recent time. The species involved may be unknown; it may be unclear whether some population represents a separate species or not; the individuals may be too small to be noticed without special sampling procedures; and the entire process of decline and extinction may extend over much more than an average human lifespan.

Positive evidence of extinction (i.e. direct observation of the death of the known last individuals) is unlikely to be available; typically, negative evidence (i.e. failure to find the species despite repeated searches) accumulates to the point where extinction is the most probable explanation. In other words, unless circumstances are exceptional, monitoring of recent extinction events has a resolution limit measured in decades, and it is thus impossible to state with precision how many species have gone extinct in any given period or to predict exactly how many species are going to become extinct by some point in the future.

From the imperfect evidence that is available, it appears that around 300-350 vertebrates and nearly 400 invertebrates have become extinct during the past 400 years. The number of plant extinctions is thought to be in the hundreds, although some believed to be extinct in the wild have survived in botanic gardens and seed collections. Because mammals and birds tend to be relatively well recorded, and leave recognisable skeletal remains, it is principally among these groups that known extinctions may be reasonably representative of actual extinctions. In these two groups the known rate of extinction over the past 400 years averages around 20-25 species per hundred years.

The key question then is: how does this compare with the average "normal" extinction rate indicated in the fossil record? Extinction rates have varied greatly, and species that are rare or otherwise prone to extinction must be poorly represented by fossils and so bias the record, but the average lifespan of a fossil species appears to be about four million years. Given this average, if 10 million species existed at any one time, the extinction rate would have been about 2.5 species annually. Applying this factor to recent birds and mammals the expected background extinction rate would be around one species every four hundred years and eight hundred years, respectively.

The known recent extinction rate appears to be some 100 or 200 times higher than background. Bias inherent in the fossil record makes it difficult to achieve greater precision in such estimates, but the general direction of the trend is well supported. Because scientific knowledge of the world's species is incomplete and highly vertebrate-centred, it is virtually certain that more extinctions are occurring than currently known. Most predictions of the contemporary extinction rate are based on combining estimates of species richness in tropical forest with estimates of rate of loss of these forests; species extinction is then predicted on the basis of the general species-area relationship, under which species richness will decline as area declines.

Projections of this sort suggest the contemporary extinction rate is very high. On a direct numerical basis, most extinctions predicted by calculations based on forest area reduction should involve beetles, because these species make up the great majority of all species in tropical forests. As a cautionary note, it should be observed that very few extinctions have to date been recorded in continental tropical moist forests, although monitoring species in these habitats presents great difficulty. Most known animal and plant extinctions have occurred on islands, and most known continental extinctions have been among freshwater organisms.

From the incomplete information available on the timing of extinction, it appears that the extinction rate (in molluscs, birds and mammals) has risen overall since about 1600 AD to near the middle of the past century and declined thereafter. The apparent decline after mid-century is probably caused in part by the time lag inherent in recording extinction, and in part by the conservation measures that many countries have taken during the latter half of the 20th century. It could also be due to the fact that extinction-prone species in the well-known groups have now been lost.

TERRESTRIAL DIVERSITY

Plant Diversity

Asia and the Pacific has a wealth of over 165 000 vascular plant species. (Figure 1) Forests in the tropical part of the region account for a major share of these species, although the Hindu Kush Himalayan belt also has as many as 25 000 plant species, comprising 10 per cent of the world's flora. Amongst subregions, Southeast Asia has the highest plant diversity, primarily because of its tropical forests. Amongst nations, Indonesia has the single highest plant diversity in the region, estimated at 37 000, and is also in the top five countries worldwide.

The country is also first in the world in terms of palm diversity, with 477 species, and one of its provinces, Irian Jaya, is home to the very rare "birds of-paradise" plant, which is only found elsewhere in the Moluccas and Australia. Australia has far and away the highest rate of endemism in the region, at 92 per cent, although high rates of endemism are also reported for the Hindu Kush Himalayan area, where for instance, of the 9 000 plant species found in the Eastern Himalayas, 3 500 species are endemic to the region.

Likewise, the Indian Himalayas contain more than 50 per cent of India's endemic flora. It is not known how many species of plants in Asia and the Pacific have already become extinct as a consequence of human activities, however, according to the IUCN Red List, more than 10 000 existing plant species in the region are threatened. Amongst subregions, Southeast Asia has the largest proportion of threatened plants.

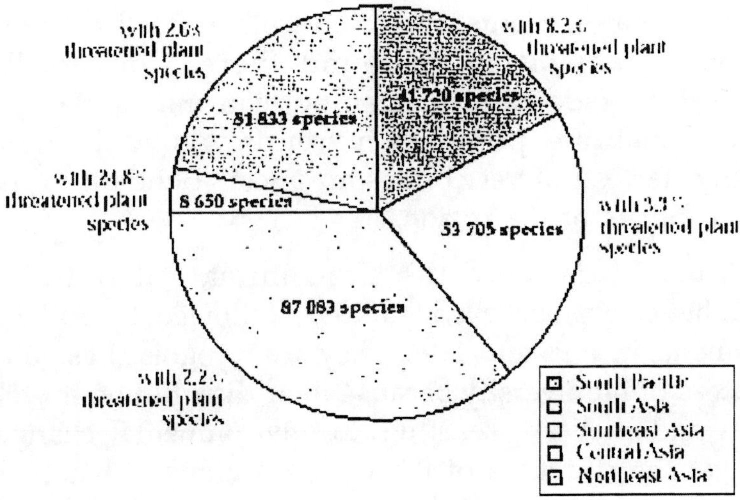

Figure 1. Total Number of Plant Species in Asia and the Pacific by Subregion and the Percentage of Threatened Plant Species

Amongst countries of the region, Turkey has the highest percentage of threatened plant species estimated at 21 per cent, closely followed by French Polynesia, at approximately 20 per cent. In terms of actual numbers of threatened species, Australia and Turkey have the highest estimates, at 2 245 and 1 876, respectively. More than 1

200 species of vascular plants in India, 707 in Japan, and about 500 each in Malaysia and New Caledonia are also threatened.

Animal Diversity

The mega diversity countries identified earlier also possess a large proportion of the animal diversity of Asia and the Pacific, a large number of which are endemic species. Vertebrates provide a good indicator of species diversity, since they generally occupy the top rungs in food chains, i.e. habitats healthy enough to maintain a full complement of native vertebrates will have a good chance of retaining the invertebrates, plants, fungi, and other small or more obscure organisms to be found there. Conversely, ecological degradation can often be read most clearly in native vertebrate population trends. As with plant diversity, terrestrial vertebrates are most abundant in the tropical forests of the region.

Birds: Birds were the first animals that IUCN comprehensively surveyed, in 1992, followed by full re-assessments in 1994 and 1996. They are recognised as good indicators of biodiversity because their distribution is well known and they are sensitive to environmental change. Amongst the countries of the region, Indonesia has both the largest number of bird species, equating to about 17 per cent of the world's total, and also one of the highest rates of endemism in the region. Endemic bird areas that are of highest priority for conservation in the region are located in the Lesser Sundas in Indonesia, Eastern Himalayas, and Luzon in the Philippines. Close to 1 100 of the region's bird species are threatened, of which 40 per cent are in the Southeast Asian subregion, and 104 are in Indonesia alone.

The most threatened major groups include rails and cranes, parrots, terrestrial game birds, and pelagic seabirds.

About one quarter of the species in each of these groups is currently threatened. Only 9 per cent of songbirds are threatened, but they still contribute the single largest group of threatened species because they are far and away the most species-rich. Other threatened bird species of the region include the crested ibis, a wading bird that has been eliminated from its former range in Japan, the Korean peninsula, and the Russian Federation, and is now down to one small population in the remote Qinling mountains of China, the Philippine monkey-eating eagle and the blue-winged Racquet-tail parrot. The relatively high numbers of threatened species in the island environments are in part a reflection of the fact that island birds tend naturally to have smaller ranges and numbers, making them more susceptible to habitat disturbance.

In the Philippines a number of bird species are believed to have disappeared due to habitat loss as a result of deforestation from the islands of Cebu, Negros, Panay and Mindoro. In Indonesia, the endemic bird habitats of the islands of Sumbu, Banggai and Sula are also under pressure. Another threat to bird species in the region, and in particular to the small island states of the Pacific, is due to predators and competition from introduced domestic animals.

Mammals: Asia and the Pacific has the largest number of mammal species in the world. Amongst the countries of the region, Indonesia has the largest number of mammal species, and also one of the highest rates of endemism in the region. Almost 1 000 of the region's mammal species are currently threatened, of which 40 per cent are in Southeast Asia. Moreover, Indonesia is also home to the orang-utan, whose populations decreased to such low levels in the early 1990s, when more than half were reportedly lost, that they were listed by the Convention on International Trade in Endangered Species of Fauna and Flora (CITES) as the most endangered species in the world.

Current estimates suggest that less than 30 000 may still be surviving in Borneo, however, this figure could even be lower since these estimates were made before fires devastated millions of hectares of forests in Borneo in the late 1990s. Other threatened species include the tiger, the largest of all cats that once ranged from Turkey to Bali and the Russian Far East. Wild tigers now barely total 3 000 to 5 000 individuals, many in small, isolated populations that are under severe threat. Species such as Caspian tiger (Panthera tigris virgata) and Near Eastern leopard (Panthera pardus ciscaucasica) have already disappeared this century from Uzbekistan.

The tiger is also regarded as extinct in the Republic of Korea, where the Siberian leopard, the fox wolf and sika are also no longer observed. Other endangered species in the Republic of Korea include the musk deer, otter and Eurasian flying squirrel. Another critically endangered specie in the region is the Bulmer's fruit bat, found only in Papua New Guinea.

Amphibians: Asia and the Pacific is home to a large number of amphibian species. Amongst the countries of the region, People's Republic of China has the largest number of amphibian species, and also the highest number of endemic species in the region. The Philippines, although comparatively low in amphibian species has high rate of endemism. The three introduced amphibian species are Bufo marinus, a marine toad introduced in the 1930s to control beetle infestation of sugarcane; Rana catesbeiana, a bullfrog introduced in the 1970s for breeding and export as food; and Rana rugulosa, introduced in the 1990s for breeding and export as food.

Of the 50 species of amphibian which are currently threatened in Asia and the Pacific, over 50 per cent are in South Pacific, and almost all of these are in Australia alone. There are also a high number of threatened amphibian species in Japan, which accounts for some 20 per cent of the overall total for the region.

Reptiles: Asia and the Pacific is home to a large number of reptile species. Amongst the countries of the region, Australia has the largest number of reptile species, and also has by far the highest rate of endemism in the region. Over 300 of the region's mammal species are currently threatened, of which over 65 per cent are in the Southeast Asian and South Pacific subregions; Australia, Myanmar and Indonesia having the highest individual totals. Among reptiles, the status of turtles, crocodilians, and tuataras has been comprehensively surveyed, but most snakes and lizards remain unassessed.

Studies aside, however, a glaring case of reptile extinction is the sea turtle whose population has declined by more than 50 per cent worldwide, largely as a result of the overexploitation of their eggs. In some localities of Southeast Asia, such as Khram Island in Thailand, only 50 green turtles and 10 hawksbills are now found compared to 158 and 45, respectively, in the 1950s. In Malaysia, the leatherback turtle has been identified as the most endangered turtle specie, with a 99 per cent drop in its numbers since man began hunting it several hundred years ago.

Barely 10 leatherbacks were sighted between 1993 and 1999, compared to 2 000 in the 1950s, and only 400 green turtles were spotted in Sarawak over the same period, compared to 4 500 in 1930s. Less than 100 green turtles can now be found in Ogasawara, Japan, as a result of an estimated harvesting of 1 800 adult green turtles each year.

Although less well known than their seagoing relatives, tortoise and river turtle species also are exploited intensively in certain areas of the region, to the point where populations have been greatly depleted. In addition, certain species of crocodilians suffer from overhunting and also from pollution of their environment, but this is one of the few taxonomic groups of animals whose overall fate has actually improved over the past two decades. Since 1971, seven alligator and crocodile species have been taken off IUCN's Red Data list, including Australia's huge estuarine crocodile.

Invertebrates: Substantial uncertainty exists over the relative abundance of invertebrate species in the tropical forests of Asia and the Pacific, although they probably represent the richest group of biodiversity in the region. Until recently, the relative diversity of arthropods in the tropics, as compared to the temperate zone, was expected to be similar to that of better known groups, such as vascular plants or birds. However, the relatively recent discovery of a tremendous richness of invertebrate species in the canopy of tropical forest of Sulawesi, Indonesia, suggests that the richness of arthropods in the tropics is much greater.

It has been estimated that as many as 30 million arthropod species-up to 96 per cent of the world's total for all species—may exist in tropical forests alone. Almost 600 of the region's known invertebrate species are currently threatened, of which the vast proportion are in the South Pacific. Australia alone contains almost 50 per cent of the subregion's threatened species.

Marine Diversity

Marine ecosystems harbour a myriad of life forms. The deep sea floor may contain about a million undiscovered species, and out of the 1.7 million species catalogued to date, around 0.25 million belong to the marine environment. However, the state of knowledge on marine species distribution and "hot spots" is currently poor because only about 7 per cent of the oceans have been sampled. Entirely new communities of organisms-hydrothermal vent communities—were found less than two decades ago in the marine ecosystems, and more than 200 new families or sub-families, 50 new genera, and 100 new species from these vents have already been identified.

The highest overall diversity takes place in the tropical Indo-Western Pacific, a region that includes waters off the coasts of Asia, northern Australia, and the Pacific Islands. Within this region, some of the highest levels of marine species richness are found off the coasts of the Philippines, Indonesia, and Papua New Guinea. It is possible that, as a result of its marine diversity, Indonesia rivals Brazil for the title of the most biodiverse country on Earth. Endemism in marine communities appears to be proportionately more dominant in areas surrounding isolated islands and thermal vents.

Broad distributions of marine species seemingly indicate that they are less vulnerable to extinction than their terrestrial kin. However, not all marine species may be as wide-ranging as is currently believed. For example, a recent effort to map the distribution of coral reef fish revealed that, of the 950 species whose ranges were mapped, one-third were limited to areas of less than 2 220 km^2. Although coral reefs share numerous attributes with tropical forests, their level of local species endemism is much lower.

Within the Indo-Pacific, for example, the vast majority of coral species are found throughout the region. Because coral reef species disperse readily, locally endemic species occur only on isolated oceanic islands. About 40 per cent of the world's mapped reefs are found in the Pacific, including the most extensive coral reef system in the world, the Great Barrier Reef of Australia, and many small islands in the South Pacific with extensive coral formations, such as Fiji, Guam, New Caledonia, Tonga, the French territories of the Pacific Islands and Tuvalu.

Coral reefs and associated species diversity, however, is greatest in Southeast Asia, which has a cluster of coral hotspots within the Indo-West Pacific region. Containing one-quarter of the world's mapped reefs, this subregion support more than 16 per cent of the world's estimated 19 000 species of freshwater and marine fish. Indonesia and the Philippines account for a major portion of these reefs, each are containing at least 2 500 species of fish. Waters surrounding Polynesia and portions of Indian Ocean also contain areas with high levels of reef fish diversity.

Amongst marine fish, the coral reef variety makes up one-quarter of all known species. Some 4 000 species of fish and 800 species of reef-building coral have been described to date, but the total number of species associated with reefs is estimated to be more than one million. Levels of coral reef species diversity vary within the marine ecosystems depending on location. The most species-rich reefs are found in a swath extending through Southeast Asia to the Great Barrier Reef, off northeastern Australia. More than 700 species of corals alone are found in this area.

The Great Barrier Reef, the world's largest system of coral reefs supports approximately 400 species of coral, 1 500 fishes, over 4 000 species of molluscs and populations of Indo-Pacific invertebrates, birds, turtles, dugong, whales, and dolphins. In addition, 252 species of bird's nest and breed on coral cays, five species of turtles live on the reef, and several species of whales and dolphins are associated with the habitat. Like their terrestrial counterpart, marine species have also been subject to pressures. About 68 per cent of all threatened marine species suffer from over-exploitation. By 1994, 90 marine mammal species were listed as threatened or endangered. More than a quarter of the world's reefs are at high risk from human disturbance. In Asia and the Pacific, coral reefs are most threatened in the Southeast Asian subregion.

FRESHWATER DIVERSITY

The overall number of species in freshwater is low compared with marine and terrestrial groups. However, species richness in relation to habitat extent is extremely high, since the area occupied by freshwater is so much lower than land or sea. As a consequence, the importance of protecting freshwater habitats is correspondingly high. Although most freshwater plant species are relatively cosmopolitan, the tropical regions of Asia appear to be one of the most rich in freshwater plant species which are restricted to a single continent, country or area. Sri Lanka, India, Myanmar and Indonesia hold localised species such as the Podostemaceae and a large number of narrowly endemic species in some cases, with several forms restricted to different stretches of a single river.

Animal species are considerably more diverse and numerous in inland waters than plants. More than half of all freshwater vertebrate species are fish, and more than 8 500 species (40 per cent) of the 25 000 known fish species exist in freshwater. Unfortunately, their distribution and systematic are inadequately known. Among the countries of Asia and the Pacific, IUCN has listed the highest number of freshwater species in Australia, followed by Papua New Guinea and Turkey (Table 1). Scientists believe that Thailand may have as many as 1 000 species of freshwater fish but only some 475 have actually been recorded. Apart from fish, other important groups of inland water species in the region include crustacea, molluscs and insects.

Table 1. Total Numbers of Threatened Freshwater Fishes in Selected Countries of Asia and the Pacific

Country	*Total species*	*Threatened species*	*Percentage of threatened species*
Australia	216	27	13
Papua New Guinea	195	12	6
Turkey	174	18	11
Japan	150	7	4
Sri Lanka	90	8	9

Accurate data required to evaluate the extinction of aquatic species are generally not available, although some 81 fish species are recorded as having become extinct during the past century, and a further 11 are extinct in the wild, but remain as captive populations. A recent survey in Malaysia found fewer than half of the 266 fish species previously known to exist in the country. Similarly, in Singapore, 18 out of 53 species of freshwater fish collected in 1934 could not be located some 30 years later, despite an exhaustive search. In the freshwater and marine ecosystems, overfishing has also resulted in shifts in fish size, abundance and species composition. For example, some 20 species of edible fish were thriving in the Aral Sea during the 1940s, but in the early 1990s, only five species remained. At present, only one is thought to survive.

Threatened Species

Various national and international programmes have

developed methods to assess the relative severity of risks faced by living species, and to label species with an indicative category name. Conservation activities can then be prioritised on the basis of relative risk, taking account of other relevant factors, such as feasibility, cost and benefits, as appropriate. The system developed by IUCN-The World Conservation Union and collaborators in conjunction with its Red Data Book and Red List programme provides a standard at the global level. To be classified as threatened with extinction, a species is assessed against a set of five quantitative criteria that form the heart of the system. These criteria are based on biological factors related to extinction risk and include: rate of decline, population size, area of geographic distribution, and degree of population and distribution fragmentation.

In general, small isolated populations will be more sensitive than larger connected ones to demographic factors or environmental factors. Human activities tend to promote fragmentation of natural and often species-rich habitats, and the spread of highly managed species-poor habitats. As a result, many species occur in just the kind of fragmented pattern that increases the risk of extinction. The conservation status of most species is not known in detail, and this certainly applies to the many million as yet undescribed species, but two large animal groups—the mammals and birds—have been comprehensively assessed.

Table 2. Threatened and extinct species

	Number of species in group	*Approx. proportion of group assessed*	*Threatened species in 2000*	*% of total in group threatened*	*Extinct species*
Vertebrates					
Mammals	4,763	100%	1,130	24%	87
Birds	9,946	100%	1,183	12%	131

Reptiles	7,970	<15%	296	4%	22
Amphibians	4,950	<15%	146	3%	5
Fishes	25,000	<10%	752	3%	92
Invertebrates					
Insects	950,000	<0.1%	555	0.06%	73
Molluscs	70,000	<5%	938	1%	303
Crustaceans	40,000	<5%	408	1%	9
Others >	100,000	<0.1%	27	0.02%	4
Plants					
Mosses	15,000	<1%	80	0.5%	3
Conifers, cycads, etc.	876	72%	141	16%	1
Flowering plants	138,000	<9%	5,390	3.5%	86

Approximately 24% (1,130) of the world's mammals and 12% (1,183) of the world's bird species are regarded on the basis of IUCN criteria as globally threatened. Proportions are much lower in other vertebrates, but none of these has been assessed fully. Empirical observations such as these give grounds for serious concern for biodiversity maintenance, regardless of any hypotheses that have been proposed regarding the contemporary and future rate of extinction.

4

Ecosystem Diversity

The defining feature of the planet Earth is that it supports living organisms, and the entire space occupied by such organisms is termed the biosphere. The living organisms in the biosphere are organised in discrete groups. Those that reproduce sexually typically exist as species, i.e. distinctive groups of similar populations that are isolated reproductively from other such groups.

Bacteria and many plants spread and reproduce vegetatively, i.e. without sexual reproduction, and the classic species concept is difficult to apply in such cases. The diversity of species, broadly defined, is nevertheless a useful general measure of the biodiversity of an area, country or the world. Globally, around 1.75 million species have been described and formally named to date, and there are good grounds for believing that several million more species exist but remain undiscovered and undescribed (Table 1).

Nearly three quarters (71%) of the Earth's surface is covered by marine waters. These have an average depth of 3.8 km, and the whole of this region, comprising virtually all the water on the planet, is theoretically capable of supporting active life. Oceans and seas thus make up the vast majority of the volume of the biosphere and by far the most extensive, if most poorly known, main ecosystem type. However, the amount of living material in most of the sea, i.e. that part of the open ocean below the upper hundred or so metres, is low compared with many terrestrial habitats.

Only two to three percent of the total world water volume is non-saline: around two thirds of this is locked away as ice and around on -third is groundwater in the upper layers of the Earth's crust. Surface freshwaters, i.e. the world's lakes, rivers and wetlands, hold the vanishingly small volume of water remaining, but this supports an important sector of global biodiversity. For example, about 40% of the more than 25,000 fish species known in the world occur in freshwaters, and many isolated water systems, large old lakes in particular, have a large number of species found nowhere else.

Land, bearing the wide diversity of terrestrial ecosystems that humans are most familiar with, as well as surface freshwaters, covers less than one third (29%) of the Earth's surface. About half of this is below 500 m elevation and the global average elevation is only about 800 m. Most of the world land surface is situated in the northern hemisphere, and the amount north of the Tropic of Cancer slightly exceeds that in the rest of the world combined.

Table 1. Estimated numbers of described species, and possible global total

Kingdoms	*Described species*	*Estimated total species*
Bacteria	4,000	1,000,000
Protoctists	80,000	600,000
Animals	1,320,000	10,600,000
Fungi	70,000	1,500,000
Plants	270,000	300,000
Total	1,744,000	ca.14,000,000

A familiar but important feature of biodiversity is that species are not evenly distributed over the planet. Although the information available on the distribution of the world's species is uneven and incomplete, the single most obvious pattern in global biodiversity is that overall species richness tends to increase toward the equator. At its simplest, this means that there are more species in total and per unit area in the tropics than in temperate regions, and more here than in polar regions. This variation in species number is strongly correlated with global variation in incident energy and water availability. These factors potentially lead to increased net primary production (NPP) by photosynthetic organisms, and a possible explanation for variation in species number is that this broader resource base may allow more species to coexist.

Marine and Coastal Ecosystem Diversity

The area of the ocean most influenced by human activity, and best known scientifically, is the continental shelf—the area of shallow water lying around the major landmasses that may be anything from a few kilometres to several hundred kilometres wide. The most landward part of this is the intertidal or littoral zone, where the bottom is periodically exposed to air and water depth varies from zero to several metres. Seaward of this the shelf slopes gently from shore to depths of one hundred to several hundred metres, forming the sublittoral or shelf zone.

At the outer edge of the shelf there is an abrupt steepening of the sea bottom, forming the continental slope, which descends to depths of 3-5 km. At this level there are immense abyssal plains, which form the floor of much of the world's oceans. The plains are punctuated by numerous submarine ridges and seamounts, which may break the sea surface to form islands. The deepest parts of the ocean are ocean trenches, which are seismically highly active, and reach depths of from 7,000 to 11,000 metres.

In the sea, as on land, photosynthesis based on sunlight is the driving force behind the maintenance of life. Because seawater absorbs sunlight strongly, photosynthesis is limited to the topmost layers of the sea, the so-called euphotic zone, which rarely reaches depths of more than 200 m in the open ocean, although at depths between 200 m and one kilometre blue light may still penetrate sufficiently to allow limited photosynthesis. Primary producers, largely in the form of photosynthesising bacteria and algae, are effectively confined to this zone.

The few exceptions include bacteria living around hydrothermal vents associated with rift zones in the ocean floor. These bacteria use hydrogen sulphide as an energy source and support a unique community of other organisms. They are scientifically of enormous interest, but make a negligible contribution to overall productivity in the oceans. Areas of high productivity are associated with upwelling zones, where nutrient-rich bottom water is brought to the surface.

Mangroves or mangals are truly hybrid terrestrial/marine ecosystems. They are a diverse collection of shrubs and trees which live in or adjacent to the intertidal zone and are thus unusual amongst vascular plants in that they are adapted to having their roots at least periodically submerged in sea water. Mangrove communities are predominantly tropical and are rarely found beyond the latitudes 30°N and 30°S. They are only able to grow on shores that are sheltered from wave action. Diversity in mangrove ecosystems is usually relatively low, particularly when compared with other tropical ecosystems such as tropical moist forests and coral reefs.

The most diverse mangrove systems are found in Southeast Asia. Coral reefs are calcium carbonate structures developed very largely by stony corals in the order Scleractinia of the phylum Cnidaria. They are essentially tropical, shallow water ecosystems largely restricted to the area between latitudes 30°N and 30°S and most abundant in shallow, well flushed marine environments characterised by clear, warm, low-nutrient waters that are of average oceanic salinity. It is difficult to measure the global area of coral reefs, although it is believed to exceed 600,000 km^2.

Near-surface reefs are estimated to cover around 255,000 km^2. They are among the most productive and diverse of all natural ecosystems, although it appears that their diversity is generally lower than that of the tropical moist forests often regarded as their terrestrial analogues. As with mangroves, the centre of diversity of reef-building corals is Southeast Asia, with an estimated minimum of 450 species of corals found associated with reefs around the Philippines, Borneo, Sulawesi and associated islands. Seagrasses are flowering plants that are adapted to live submerged in seawater.

There are approximately 48 species found in shallow coastal areas between the Arctic and Antarctic, down to depths of around 60 m. The most extensive areas are on soft substrates. Seagrass beds have high productivity and contribute significantly to the total primary production of inshore waters, being an important source of food for many organisms. The total global area is unknown. Algae are the major large photosynthesising organisms in marine environments. There are three main groups: the red algae or Rhodophyta with around 4,000 marine species, the green algae or Chlorophyta with around 1,000 species, and the brown algae or Phaeophyta with about 1,500.

In many of the colder regions of the world, hard subtidal substrates are occupied by very large brown algae collectively known as kelps, which form extensive kelp beds or forests. The cooler regions of the world are rich in algal species. Pelagic or open-ocean communities occupy a greater area than any other major community type on Earth. They are dominated by the activity of plankton in the euphotic zone near the surface. The plankton support a large number of free-swimming organisms or nekton, predominantly fishes but also cetaceans and cephalopod molluscs.

The marked vertical gradients within the pelagic zone—of light, temperature, pressure, nutrient availability and salinity lead to strong vertical structuring of pelagic species assemblages. This vertical stratification is not static, but fluctuates seasonally and, often, daily, with variations in the physical and chemical characteristics of the seawater and vertical migrations of mid-water organisms. Until recently it was assumed that biomass in the pelagic zone everywhere below the euphotic zone was low.

Recent studies have indicated that in some areas, biomass of tropical mid-water or mesopelagic animals may be surprisingly high. Deep-sea communities are prevalent on around half of the world's surface area, where the ocean is over 3,000 m deep. All such areas are in permanent darkness. Biomass is very low but new sampling techniques indicate that the diversity of small organisms in or on the sea bottom is often relatively high.

The world ocean plays a crucial role in the regulation of climate and in the carbon cycle and other biogeochemical cycles. Much of this role can be ascribed to the ocean as a physical entity, that is as a vast, circulating body of water whose high specific heat capacity buffers atmospheric changes in temperature and which is capable of dissolving very large amounts of carbon dioxide. It does seem however that marine organisms also play a very important role in the major biogeochemical cycles, although one which is difficult to quantify.

Marine fisheries provide by far the largest source of wild protein, this being of critical importance to many subsistence communities around the world, making use of fin fishes, crustaceans and molluscs, with marine algae increasing in importance in Asia. Mangroves and seagrass beds fulfil an important function as nursery areas for juvenile fish and shellfish. Marine capture fisheries have increased in volume nearly five-fold in the past 50 years, rising to nearly 90 million tonnes in the late 1990s, this making up more than 70% of the total world production of aquatic resources. Analysis by FAO and others indicates that marine stocks are widely overexploited and in urgent need of remedial management. Other major uses of the marine sphere include:

— waste disposal;

— recreation, chiefly in coastal zones;

— coastal stabilisation;

— transportation.

Impacts on Marine and Coastal Ecosystems

Impacts can be grouped into five main categories:

— chemical pollution and eutrophication;

— fisheries operations;

— global climate change;

— alterations of physical habitat;

— invasions of exotic species.

Chemical pollution with eutrophication is a widespread problem, most pronounced in semi-enclosed seas such as the Mediterranean and the Black Sea. In addition to the potential for over-fishing, fishery operations can have a destructive physical impact on the seabed, and affect population levels of non-target species through incidental catch, such problems being of particular significance for cetaceans, sea turtles and seabirds such as albatross, in different parts of the world. All commercial bottom fishing disturbs sea-floor organisms and the seabed, with impacts on both habitats and species.

Coral bleaching related to periodic climatic events is the primary threat to coral reefs on the global scale. The increase in sea-surface temperature associated with the major El Niño and La Niña climate switches in 1997-1998 resulted in extensive coral bleaching and mortality over large portions of the Indian Ocean and Southeast and East Asia. On some reefs, there were mortality levels greater than 90% leaving some reefs almost bare of corals and with early indication of major shifts in the population structures.

The critical feature of recent coral-bleaching events is that areas have been struck indiscriminately, irrespective of the status of reef health; impacts have been felt both on pristine, remote reefs and on reefs already under major human-induced stresses. Some of the changes recently caused by periodic climate events, including coral bleaching, are not necessarily permanent, however human-induced stresses causing physical degradation and destruction to coral-reef organisms often exacerbate the effects of these events or limit the recovery capability of reef ecosystems.

Status and Trends of Species

As a gross generalisation, marine species appear to be somewhat less prone to extinction than inland water or terrestrial ones. Certainly, available information indicates that far few marine species are known to have become extinct since 1600 than either terrestrial or freshwater ones. Part of this difference may be because marine species are in general much less readily observable than those in terrestrial habitats, and, to a lesser extent, inland water species. It is possible therefore that a higher proportion of marine than terrestrial or inland water species has become extinct without our knowing. However, it seems likely that the difference is in large part a real one, and can be ascribed to two main factors.

First, because of the size of the oceans, and the fact that people do not live permanently in them, they are as a whole considerably more buffered from human impacts than terrestrial or inland water areas.

Second, marine species appear on the whole to be more widespread than terrestrial or inland water ones. This makes them generally less vulnerable to extinction.

Inland Water Ecosystem Biodiversity

Freshwaters make up less than 3% of the total volume of water on Earth. They consist largely of water in the form of polar ice and groundwater, i.e. water below the Earth's surface held within rocks or between rock strata and constituting perhaps 30% of global freshwater resources. Water in lakes and rivers constitutes less than one hundredth of one percent (<0.01) of the world's total water volume, and around a quarter of one percent (<0.27) of global freshwater volume. Lake water is the largest component in this vanishingly small subtotal.

The more superficial groundwater deposits are linked to the global water cycle, and are used for human consumption or agricultural purposes, whereas the deeper layers tend to be somewhat saline and do not participate in exchanges with other parts of the system. Inland aquatic habitats show far more variety in their physical and chemical characteristics than marine habitats. They range from vast, permanent lakes and rivers to geothermal springs, temporary puddles and minute pools and from almost pure water to highly concentrated solutions of mineral salts, toxic to all but a few specialised organisms. Inland aquatic habitats can be divided into running or lotic and standing or lentic systems. They may also be divided into permanent water bodies, periodically inundated and transient or ephemeral.

There is no rigid dividing line between an inland aquatic habitat on the one hand and a terrestrial or marine habitat on the other. Seasonally inundated areas, such as river floodplains, are effectively hybrid or transitional systems. There are also many areas that consist of shifting mosaics of land and shallow water, or areas of saturated vegetation, such as sphagnum moss bogs, that are neither strictly land nor water. There areas are often termed "wetlands."

Rivers and other lotic systems are essentially linear bodies of water draining under the influence of gravity from elevated areas of land toward sea level. River systems typically branch into ever smaller and more numerous channels with increasing elevation. The volume of water and speed of flow usually varies greatly from one part of a river system to another. These factors also vary seasonally; in some arid areas rivers flow for only part of the year, or in extreme cases only once in several years.

Variations in water flow and in underlying geology create a wide range of habitats within any river, often within a short distance. Large river systems may also span many degrees of latitude and pass through a wide range of climatic conditions within their catchments. This variation creates a wide range of niches for different organisms within any given river system. Although river systems are constantly changing their courses, sometimes radically, through deposition and erosion of channels and the uplift and erosion of watershed uplands, evidence suggests that many larger river systems are extremely old, with some having been in continuous existence for tens of millions of years.

Lakes, of which around 10,000 exceed 1 km^2 in extent, are mainly glacial in origin and geologically very young, dating from the retreat of the continental ice-sheets at the start of the Holocene, some 11,500 years before present. All such lakes are expected to fill slowly with sediment and plant biomass and to disappear within perhaps the next 100,000 years, along with any isolated biota. Most other lakes are caused by tectonic activity, either through faults caused by deep crustal activity or through vulcanism. Lakes may also be formed by dissolution of soluble rocks or by changes in the course of rivers in floodplain regions.

Only around ten existing lakes are known with certainty to have origins much before the Holocene. Most of these occupy basins formed by large-scale subsidence of the Earth's crust, dating back to at most 20 million or 30 million years before present. Wetlands are traditionally understood as heterogeneous habitats of permanent or seasonal shallow water dominated by large aquatic plants and broken into diverse microhabitats. The four major habitat types are:

— Bogs, which are peat-producing wetlands in moist climates where organic matter has accumulated over long periods. Water and nutrient input is entirely through precipitation. Bogs are typically acid and deficient in nutrients and are often dominated by sphagnum moss. They are also important habitats for carnivorous plants. In many places they are endangered by the extraction of peat for horticulture.

— Fens are peat-producing wetlands that are influenced by soil nutrients flowing through the system and are typically supplied by mineral-rich groundwater. They are usually more productive and less acidic than bogs.

— Marshes are inundated areas with herbaceous emergent vegetation, commonly dominated by grasses, sedges or reeds. They may be either permanent or seasonal and are fed by ground or river water, or both.

— Swamps are forested wetlands on waterlogged or inundated soils where little or no peat accumulation occurs. Like marshes they may be either seasonal or permanent.

At higher taxonomic levels, the diversity of organisms in inland waters is considerably lower than on land or in the sea. The number of species overall is also low compared with marine or terrestrial habitats; however, species richness compared to habitat extent may be very high. For example, of known fish species around 40% are freshwater forms and 60% marine, despite the fact there is around 10,000 times as much seawater as there is water in inland water habitats. This high diversity of inland water fishes relative to habitat extent is undoubtedly promoted by the extent of isolation between inland water systems.

As is the case with terrestrial habitats, species richness overall increases strongly toward the equator, so that there are far more species in tropical than in temperate or polar regions. Among river systems, the Amazon-Ucayali is both by far the largest river catchment in the world, holding the greatest volume of water, and is also wholly within the tropics. Unsurprisingly, it is the major repository of the world's inland water biodiversity, although it remains very incompletely catalogued. Among lentic systems, the most diverse are the large African rift valley lakes, notably Tanganyika, Victoria and Malawi.

Inland waters have in many areas suffered as a result of the conflicting interests of different sectors. Physical change to freshwater habitats is a prime cause of decline and extinction of fish species, but may provide benefits, such as hydroelectric power, to human populations. Dam construction is the chief cause of extinction in the formerly large gastropod fauna in Mobile Bay (USA), for example. The introduction of alien species is a second prime cause of decline: the impact of Nile Perch on the native cichlid fauna in Lake Victoria is well documented.

Overall, the major kinds of impact on inland waters are:

— physical alteration and destruction of habitat through abstraction of water, drainage, canalisation, and flood-control systems;

— construction of dams and reservoirs;

— sedimentation;

— introduced species;

— pollution, including: eutrophication, acid deposition, salinisation, heavy metals.

Status and Trends of Species

The status of inland water faunas has been considerably worse than originally suspected. Of the 20 or so countries that have been reasonably comprehensively assessed, on average just under 20% of the inland water fish fauna has been found to be threatened. The proportion of inland water chelonians (tortoises and turtles) threatened is even higher, with more than 30% of species having been assessed as critically endangered, endangered or vulnerable. Amongst birds and mammals the proportions are considerably lower, probably because many semi-aquatic species are able to disperse relatively easily from one inland water body to another.

A common finding among global assessments of the status and trends of biological diversity is that inland waters are those suffering the greatest deleterious impact from human activities at present.

Forest Ecosystem Biodiversity

Forests are ecosystems in which trees are the predominant life forms. A more precise definition than this remains surprisingly elusive, because trees occur in many different ecosystems, at different densities and in different forms. Most definitions of forest refer to canopy or crown cover, which is essentially the percentage of ground area shaded by the crowns of the trees when they are in full leaf. Clearly, estimates of the area of forest both globally and in any given place must vary enormously depending on the definition of forest adopted.

The factors determining the distribution of forests are largely climatic: tree establishment and growth requires a minimum number of days in the year with adequate moisture and warmth for active growth. Substrate characteristics are also important as trees require access to enough soil for supply of nutrients and water, and to provide anchorage. Trees are therefore absent from areas that are too dry, or too cold or have inadequate soil cover.

Other natural factors that may limit the distribution of forests include fire, flooding, the presence of toxic minerals in the substrate, and the impact of large herbivores. Around one half of the Earth's land surface is climatically suitable for forest, but as a result of human actions, around one half of this area is not forested, so that current forest cover amounts to just under 40 million square kilometres.

Classifying forests is in some ways an even more difficult task than defining forest. A number of global classification systems have been proposed, some extremely complex, but as yet none has gained universal acceptance. All such systems are more or less artificial as in nature forests rarely form homogenous units clearly separable from other such units; instead species composition and forest structure tend to change gradually or irregularly with, for example, changing altitude, exposure and latitude.

Boreal and temperate needleleaf forests cover a larger area of the world than any other forest type. They occur mostly in the higher latitudes of the northern hemisphere as well as high altitude zones and some warm temperate areas. Conifer canopies are efficient light absorbers, so limiting opportunity for extensive development of lower strata of vegetation, and the structure of these forests is often fairly simple. Species diversity is usually relatively low, and vast expanses of such forest in the northern hemisphere are dominated by a very small number of tree species.

However, diversity amongst some groups of organisms, such as mosses and lichens, may be surprisingly high. These forests are of great importance in the carbon cycle, acting as major reservoirs of organic carbon both above and below ground. Temperate broadleaf and mixed forests are generally characteristic of the warmer temperate latitudes but extend to cool temperate ones, particularly in the southern hemisphere. These forests tend to be structurally more complex than needleleaf forests, and to have considerably higher species diversity.

Aboveground biomass tends to be lower than in temperate needleaf or tropical moist forests, while that below ground tends to be intermediate. Tropical moist forests include many different forest types. They are without doubt overall the most diverse ecosystems on earth. Some estimates suggest that at least 60% of all species occur in them, despite the fact that they cover little more than 7% of the world's land surface, and around 2% of the surface of the globe.

Diversity is usually extremely high at all spatial scales; in some parts of the western Amazon and of the Atlantic coastal forest of Brazil there may be as many as 300 tree species per hectare. Notable exceptions are mangrove ecosystems, which have low tree species diversity, and forests on very nutrient-poor soils. Diversity also tends to decrease with increasing altitude and with increasing seasonality of climate. Lowland tropical moist forests can have very high aboveground biomass, although not as high as some needleleaf forests. Belowground biomass, however, with the exception of some peat-swamp forests, tends to be relatively low.

Tropical moist forests are estimated to account for nearly one third of global terrestrial annual net primary production. Tropical dry forests are characteristic of areas in the tropics affected by seasonal drought. Seasonality of rainfall often results in a largely deciduous forest canopy; however, under some conditions, such a low fertility soils or unpredictable rainfall regimes, the canopy may become dominated by evergreen sclerophyllous species. On very poor soils and especially where fire is a recurrent phenomenon, woody savannas may develop.

In general dry tropical forests have lower species diversity than tropical moist forests; however they appear to be characterised by high levels of local endemism. Sparse trees and parkland are forests with open canopies of 10-30% crown cover. They occur principally in areas of transition from forested to non-forested landscapes. The two major zones where these formations occur are the boreal region (i.e. at very high latitudes) and the seasonally dry tropics.

Forest plantations, generally intended for the production of timber and pulpwood, are believed to cover well over one million square kilometres worldwide. Commonly composed of only one species, often non-native, plantations are not generally important as habitat for native species. However, they can be managed in ways that enhance the role they play in maintaining biodiversity. They may also be important providers of ecosystem services, such as maintaining nutrient capital and soil structure as well as protecting watersheds. They may also alleviate pressure on natural forests for timber and fuelwood production.

The Earth's climate has been extremely unstable for the past two million years and for most of this time much colder than today's. The extent and distribution of forests during the cold periods were very different from today, not least because much of the northern hemisphere was covered in thick ice, but also because the climate in much of the tropics was apparently drier. The beginning of the Holocene, which followed the last major glacial period between 11,000 and 12,000 years ago, would undoubtedly have been a period of major forest expansion as the ice retreated and wetter conditions began to prevail in much of the tropics. However, it also evidently marked the emergence of agriculture and the spread of humans through almost all the habitable world, including the Americas.

Human impact on forest cover undoubtedly dates back at least to this period, but because these impacts started at a time when forest extent and distribution would have been rapidly changing also for climatic reasons, it is extremely difficult to determine just how much land has been permanently deforested by human hands. In other words, it is very difficult, and perhaps of little value, to seek a precise global baseline of "original forest cover" against which to measure human impact. Such a baseline could theoretically be established before the start of the Pleistocene ice ages, over two million years ago, but the climate at that time is not sufficiently well known to make this feasible.

Despite these problems, it is possible to make some general observations on the history of forest loss, which differs markedly between different regions and different forest types. In Western Europe it seems that forest clearance has an extremely long history, with, for example, much of the British Isles deforested in the Neolithic, between 7,000 and 5,000 years ago. In Europe as a whole, forest cover continued to decline until the 19th century when it stabilised. Since the early 20th century, forest cover in Europe has expanded, often through the establishment of conifer plantations although latterly also through reestablishment of mixed and broadleaf forests.

Similarly, in eastern North America, forest cover reached a minimum around 1860, but has since expanded. Forests west of the Appalachians in North America suffered the most severe impacts in the late 19th and early 20th centuries, but are still under pressure from demand for timber and pulp. It has been widely assumed that large-scale forest clearance outside of Europe had only taken place following European colonisation. However, there is increasing evidence that indigenous peoples in some areas may have had major impacts on forest cover before the arrival of Europeans. Nevertheless, in most areas European colonisation undoubtedly precipitated a major phase of deforestation, which appears to have reached unprecedentedly high levels in the last half-century or so.

Major Impacts on Forests

— Conversion, chiefly to cropland (including plantation forestry) but also for a range of other purposes, including road building, mining, urban and industrial development,

— Fragmentation,

— Changing fire regimes,

— Invasive alien species,

— Logging,

— Extraction of non-timber forest products,

— Fuelwood extraction,

— Hunting,

— Unsustainable shifting cultivation,

— Climate change,

— Pollutants, including acid rain.

Status and Trends of Species

Because forests occur over such a wide area and encompass such a wide range of ecological conditions, it is difficult to generalise about the status of forest species as a whole. Most temperate and boreal forest species in the northern hemisphere are widespread and generally not highly threatened with extinction. Forests of warmer temperate and subtropical areas have a higher diversity of species, many of which are localised and therefore in general more vulnerable to extinction.

A higher proportion of these is threatened. Similarly, southern hemisphere temperate forests are generally isolated from each other and have a high proportion of localised and threatened species. For those groups, namely mammals and birds, that have been analysed in any detail, far more tropical forest species are considered threatened than species from any other habitat.

It is unclear, however, whether this is merely a reflection of the fact that tropical forests contain far more species in these groups than any other habitat, or whether a higher proportion of these species is threatened than of temperate and boreal forest species. In tropical and subtropical regions, there are clear differences between insular and continental biotas. The former appear far more extinction-prone and have a far higher proportion of threatened species than the latter.

Dry and Sub-humid Land Ecosystem Biodiversity

All life requires water. On land the abundance and diversity of life as well as the kinds of life-form that exist in any given area are overwhelmingly influenced by two major factors: the amount of available moisture and the temperature. Variability in these factors, on daily, seasonal and inter-annual scales, is as important as long-term averages. Aridity is defined on the basis of the ratio of rainfall per unit area to the potential loss of water from that area through plant use and evaporation.

Humid areas, defined by an aridity index of 0.65 or above, extend over more than 39% of the Earth's land surface. The remaining non-humid 61% is made up of the cold regions, and the drylands. The former include polar and tundra areas, and certain high mountains and plateaus, together covering nearly 14% of the land. They differ ecologically from other non-humid areas, mainly by having temperatures below freezing for a period long enough to restrict or prevent plant growth, although they are "dry" in the sense that liquid water is unavailable for a significant part of the year.

The true drylands extend over nearly half (47%) of the Earth's land surface, and can be subdivided into a number of zones on the basis of the aridity index. The hyperarid regions have very low biological productivity, and little or no opportunity for human occupation and human-induced land degradation. The less extreme and more productive arid, semiarid and dry sub-humid areas are liable to become degraded by human activity, and are collectively known as the susceptible drylands.

Drylands contain a very wide range of natural habitats, including barren desert with virtually no visible signs of life, semi-desert dominated by succulents and other xerophytic plants, grassland, savannah, and many different kinds of scrub- or shrubland, woodlands and forests. However, in contrast to humid areas, where the dominant natural vegetation is usually forest, many drylands are characterised by sparse or absent tree cover. This may be because the climate is simply too dry to support closed forest ecosystems, or because the area is too severely affected by fire or grazing. Dry and sub-humid ecosystems are the centres of origin of many major crops.

Types of Dry and Sub-humid Lands

Mediterranean ecosystems are typified climatically by generally cool, wet winters and warm or hot dry summers. However, no single climatic or bioclimatic definition of a Mediterranean ecosystem has yet been established. Mediterranean ecosystems encompass a wide range of habitat types including forest, woodland and grassland, but are typified by a low, woody, fire-adapted sclerophyllous shrubland (variously known as maquis, chaparral, fynbos, mallee) on relatively nutrient-poor soils.

These systems occur in five distinct parts of the world: the Mediterranean basin, California (USA), central Chile, Cape Province (South Africa), and southwestern and south Australia.

Each of these regions occurs on the west side of a continent and to the east of a cold ocean current that generates winter rainfall. They cover around 2.5 million km^2 in total, or between 1% and 2% of the Earth's surface, more than three quarters of which is within the Mediterranean basin. They are disproportionately rich plant species compared with most other parts of the world. In most parts of the world where they are found, a proportion of ecosystems generally classified as Mediterranean occur within dryland areas according to the definition given above. Elsewhere, including those regions generally associated with the typical Mediterranean high diversity sclerophyllous vegetation, Mediterranean-type ecosystems occur in humid climatic zones and are therefore not classifiable as drylands.

Grassland ecosystems may be loosely defined as areas dominated by grasses or grass-like plants with few woody plants. Natural grassland ecosystems are typically characteristic of areas with three main features: periodic drought; fire; and grazing by large herbivores. In addition, they are often associated with soils of low fertility. The relative importance of different factors in maintaining grasslands varies locally and regionally. Savannas are tropical ecosystems characterised by dominance at the ground layer of grasses and grass-like plants. They form a continuum from treeless plains through open woodlands to virtually closed-canopy woodland with a grassy understorey.

Most, but by no means all, of the world's natural grasslands and savannas are found in dryland regions. Around 20% of the Earth's land surface supports grasslands of varying degrees of naturalness; temperate grasslands make up approximately one fourth of this area, and savannas the remainder. Seasonally flooded grassland areas are found in many river basins and may be of considerable ecological and biotic importance. Such areas may be considered grassland ecosystems or inland water (wetland) ecosystems; there is no clear dividing line between the two.

Major Impacts

Humans have had enormous impact on dryland, grassland and Mediterranean ecosystems, often with major negative impacts on biological diversity. These impacts are often complex and interrelated. The following are among the major categories of activities that have or can have an adverse impact:

— Conversion, particularly to cropland. Large-scale transformation to cropland in drylands is dependent on availability of freshwater, either as surface waters (lakes, rivers, reservoirs) or in aquifers. In many areas, conversion is followed by abandonment as soil becomes degraded, often through salinisation or waterlogging.

— Changing fire regimes. Many dryland ecosystems are naturally adapted to some degree to fire. However, in many areas fires caused by humans (often as deliberate burning to produce new growth for livestock) are far more frequent and extensive than natural fires, and have very different impact on ecosystems and their biodiversity.

— Impact of introduced herbivores, particularly livestock, through trampling, removal of plant biomass, introduction of pathogens and alteration of plant species composition through selective grazing and competition with native species.

— Introduction of non-native plants. Grassland ecosystems have been routinely modified by deliberate introduction of non-native species, particularly other grasses and leguminous plants.

— Water. By definition, water is a limiting resource in dryland ecosystems. Human use of existing water resources in these systems therefore has often a disproportionately extreme impact. Abstraction of water for irrigation from freshwater systems such as lakes and rivers has an often extreme impact on the biodiversity of these ecosystems. Depletion of groundwater resources may have a less obvious direct impact but is likely to affect natural ground-fed springs and deep-rooted vegetation where the aquifer is relatively near the surface. Creation of artificial water sources for livestock leads to the creation of virtually barren "sacrifice zones" around the water source as a result of extremely heavy trampling by livestock.

— One of the most significant human impacts in natural or seminatural dryland and savannah ecosystems in developing countries is through harvest of wood for fuel. Quantifying this, and assessing its long-term impact on ecosystems, has proved problematic, although in some areas the impact is undoubtedly severe. Overharvest of wild species. Excess hunting of wildlife and collection of plants, whether for subsistence use or national or international trade, can have severe impacts, in some cases driving species to extinction. Because dryland species tend to have relatively low population growth rates, and in the case of plants, individual growth rates, they may be particularly sensitive to overharvest.

— Chemical inputs. In many grassland ecosystems, highest biological diversity seems to be associated with poorer soils. Artificial enrichment of grasslands, particularly through application of nitrogenous fertiliser, generally leads to a very marked decrease in plant species diversity.

— Long-term impacts of climate change. The potential impacts on dryland ecosystems of human-induced climate change remain to be quantified but are likely to be significant.

These various impacts interact in a complex and sometimes unpredictable way. In drylands, the collective effect of factors, mainly of human origin, leading to land degradation make up a process often termed desertification. According to the United Nations Environment Programme (UNEP), desertification directly affects some 36 million km^2 of the world's drylands, i.e. about 70% of the total dryland area, and one-sixth of the world's people. The effects of desertification promote poverty among rural people, and by placing greater stress on natural resources, poverty tends to reinforce any existing trend toward desertification.

Status and Trends in Species

Summary analysis of the habitat distribution of globally threatened mammals and birds shows clearly that while the majority of species occupy forest habitats (particularly lowland tropical moist forest), dryland, scrub and grasslands made up the second most important group of habitats (being somewhat more critical for mammals than for birds, in which group wetlands were significant). There are generally insufficient data overall at present to determine whether species in the ecosystem types under discussion are relatively more or less prone to extinction than those elsewhere—that is whether a higher or lower than average proportion of species in these ecosystems can be classified as threatened.

With respect to mammals, a high percentage of continental species believed or known to have become extinct since 1600 occurred in dryland ecosystems, most notably in Australia. In general, dryland mammals tend to be relatively wide-ranging but to occur at low population densities because of the low primary productivity of these areas. Larger species are also more conspicuous than forest dwelling species and thus more vulnerable to overhunting. These factors have meant that a notable number of large dryland mammals are either highly endangered or extinct in the wild.

The Mediterranean-type ecosystems in general have a relatively high proportion of their species categorised as threatened. This is in part a consequence of human land-use development in agriculture, industry and housing, and, especially in Cape Province (South Africa) and in California, of the spread of non-native plant species. The Cape flora, largely within a Mediterranean-type ecosystem, occupies only 4% of the land area of southern Africa, but accounts for nearly 70% of the region's threatened species. Around 10% of the California flora is considered threatened.

The main threats are: urban coastal development, pollution, agriculture, tourism, water shortages and fire. Forests and woodlands, and inland water ecosystems within drylands tend to be placed under disproportionate pressure. This is chiefly because they are inherently scarce resources but also, in the case of forests and woodlands, because they tend to be found in areas within drylands with somewhat more favourable conditions (microclimate and soil fertility) than normal. These areas are therefore most likely to be settled by people and suffer extensive habitat conversion. Mediterranean ecosystems in particular are noted for their amenable climates and therefore come under particular pressure for permanent settlement and tourism.

Biodiversity of Agricultural Ecosystems

Agricultural biodiversity is the diversity at all levels of the biological hierarchy, from genes to ecosystems, that is involved in agriculture and food production. The diverse range of organisms making up agricultural biodiversity can be divided into three major groups, based on the way they contribute to or affect agricultural production. The fundamental and distinct property of agricultural biodiversity is that it is largely created, maintained and managed by humans, originally by subsistence farming communities, but more recently also by biotechnologists, in part using material in *ex situ* genetic resource collections. In this regard, agricultural biodiversity stands in total contrast to wild biodiversity, which is most valued in situ and as a product of natural evolution.

Although agricultural ecosystems provide some of the environmental services provided by wild ecosystems, in that they can, for example, protect soil structure and affect air and water quality, the magnitude of such benefits from converted ecosystems will generally not exceed those provided by natural ecosystems. The cultural and spiritual benefits that are often associated with traditional agricultural landscapes provide an important exception to this generalisation.

Wild species and their products used for food are not usually regarded as part of agricultural biodiversity unless there is some degree of resource management involved, so, for example, species involved in coastal or inland fish farming come within the scope of agricultural biodiversity, but high seas fishery species do not. The components of agricultural biodiversity variously occur in protected areas, seed banks, laboratories, and the stores of industrial seed producers, but their primary habitat is land supporting agricultural production. Although most crop production is rainfed and based in the cool to warm humid regions of the world, production is extended into drylands by irrigation.

Domestic livestock can thrive under a variety of climatic conditions. The ability of sheep, goats and camels to exist on sparse vegetation with little water allows humans to occupy marginal drylands, and the semi-domestic reindeer ranges into arctic regions. At the close of the Pleistocene, some 11,500 years before present, domestication of plants and animals had just begun and the area of agricultural land would have been imperceptibly small.

The history of agriculture is a history of experimentation with plant and animal genetic resources by human communities, and the dispersion of these resources by trade and the migration of peoples. Agricultural land now forms a significant proportion, some 38%, of the world's total land area. Clearly, at global level, the habitat of agricultural biodiversity has increased enormously during the past ten millennia. Although some care is needed in interpreting these data, a graph of estimates of agricultural land area shows the rate of increase from the 1960s to the present. An increase in area attributed to one classification unit must entail a decrease in some other land cover type; in this case, the increase in agricultural land is accompanied by a decrease in forest and woodland area, and a decrease in the category "all other land."

Replacement by modern, genetically more uniform breeds specialised for intensive systems is the main cause of loss of livestock breeds. The full significance of the loss of varieties of crop plants and livestock, and their wild relatives, is difficult to evaluate. The loss of a cultivated plant variety might entail little or no loss of individual genes not present in other varieties although a particular pattern of gene regulation or interaction may be lost. It can be assumed, on the basis of much evidence, often anecdotal, that many local varieties possessed features of adaptive value in a local context, and their continued existence presents the opportunity for using the genes responsible in breeding improvements more generally. The precautionary principle implies that such diversity should be conserved.

Soil degradation can include impairment of ecological services mediated by soil organisms, such as bacteria, fungi and small invertebrates, but there is little detailed information available on the overall status of soil biodiversity, or the resilience of these species following environmental change. There is, however, an increasing amount of information relating to declines in abundance of some pollinators. It is estimated that twothirds of the world's species of agricultural crops require animals for pollination. Species responsible for this service include birds, wasps, beetles, butterflies, bats, moths, giraffes, opossums and flies, but by far the greatest part is provided by bees.

Although most estimates of the economic value of crop pollinators give credit to the honeybee, many other species of bee are involved. However the numbers of native bees are dwindling. The losses are due mostly to the use of agrochemicals and monocultures, and to deforestation. The chemicals kill bees and, with the removal of wild vegetation and crop specialisation, the bees find neither places to nest nor alternative flowers while they wait for the crop to bloom.

In addition, honeybees in many parts of the world have contracted a serious disease and the numbers of honeybee colonies have decreased dramatically. Pollination can be a limiting factor to yields. The rapid development of transgenic crops raises additional causes for concern, as the employment of a herbicide or pesticide coupled with a variety of crop resistant to this could eradicate all alternative forage for pollinators, leading to devastation in their numbers.

Biological Diversity and Global Environmental Issues

The strong scientific linkages between global environmental issues -such as loss of biological diversity, climate change, stratospheric ozone depletion, water degradation, or the accumulation of persistent organic pollutants—are becoming increasingly apparent. The issue that may pose one of the greatest threats to biodiversity is climate change. For this reason, one of the most critical tasks is to identify the scientific and policy interlinkages between biodiversity loss and climate change.

Climate Change

The weight of scientific evidence suggests that the observed changes in climate are caused, at least in part, by human activities, primarily the burning of fossil fuels and changes in land cover. These activities are modifying the concentration of CO_2 and other greenhouse gases which absorb heat radiating from the earth as well as the properties of the surface which absorbs or scatters radiant energy. Climate change may directly affect species through changes in phenology, lengthening of the growing season, and changes in distribution.

In many cases the observed changes are consistent with well-known biological responses to climate. Changes in such characteristics of organisms may thus serve as indicators or early warnings of climate change. Climate change is an additional stress on ecosystems and species that are, often, already under stress from other pressures such as: habitat change resulting from land-use change; overharvesting; pollution; and the effects of invasive species. Such pressures thus make biodiversity more vulnerable to climate change. For example:

— Habitat fragmentation poses barriers to dispersal, thereby reducing the possibility that species can adapt by moving as the climate changes.

— Habitat fragmentation and overharvesting may result in small isolated populations with low genetic diversity. With low genetic adaptability such populations will be more vulnerable to extinction;

— Ecosystem degradation, which may result from unsustainable use of ecosystem components, pollution, pest outbreaks, or changes in fire regimes, can decrease the resilience of ecosystems to climate change.

Addressing such exacerbating factors may be an important component of adaptation to climate change. The expected result of these interactions is that climate change will lead to reduced biological diversity. At the species level, those that are most likely to be adversely affected are those that can only tolerate a narrow range of climatic conditions and that have limited capacity to disperse. This is likely to include a significant proportion of already threatened species.

At the ecosystem level, established natural communities may be broken up as the constituent species shift at different rates in response to climate change. For example, a substantial fraction of the world's forested area is expected to undergo major changes in broad vegetation types with the greatest changes at high latitudes. New assemblages of species and hence new ecosystems may be established. As noted below, this may have major implications for the role of forests as carbon stores. Differential responses to climate change by species in ecosystems may lead to disruption of important functional interactions, with potentially highly serious consequences for the provision of ecosystem services such as pest control, pollination, seed dispersal, decomposition and soil nutrient cycling.

In addition to the effects on natural ecosystems, these could have socio-economic consequences for agriculture. The impact of climate change on biological diversity is expected to be non-linear. The impact may be particularly severe when certain critical thresholds are crossed. Ecosystem types that are vulnerable to such thresholds include:

— *Wetlands overlying permafrost*: These are likely to be severely affected when the ice melts;

— *Coral reefs*: As already noted by the Conference of the Parties, there is significant evidence that climate change is a primary cause of the recent and severe extensive coral bleaching. Bleaching is reversible when the increases are short-term and of no more than 1-2°C. However, sustained increases in water temperatures of 3-4°C above normal maxima can cause significant coral mortality. Severe bleaching events were triggered, for example, by the El Niño events of 1982-83 and 1997-98.

Climate change may also increase threats from invasive alien species:

— Climate change may result in extension or changes in the ranges suitable to certain invasive species. An example may be the increased prevalence of vector-borne infectious diseases transmitted by blood-feeding mosquitoes and ticks;

— Environments may become more favourable to weedy species because of climate change induced ecosystem disruptions.

In short, and as the Intergovernmental Panel on Climate Change (IPCC) has concluded, ecosystems vital to human development and well-being are vulnerable to climate change. There are likely to be reductions in biological diversity and in the goods and services that ecosystems provide to society, e.g. sources of food, fibre, medicines, recreation and tourism, and ecological services such as controlling nutrient cycling, waste quality, water runoff, soil erosion, pollination services, detoxification and air quality. Additionally there may be an increased provision of ecosystem "bads" such as pests, diseases and other invasive species.

5

The Value of Agricultural Biodiversity

Agricultural biodiversity includes all components of biological diversity of relevance to food and agriculture. It is essential to the world for the following functions:

- sustainable production of food and other agricultural products, including providing the building blocks for the evolution or deliberate breeding of useful new crop varieties;
- biological support to production via, for example soil biota, pollinators, and predators;
- wider ecological services provided by agro-ecosystems, such as landscape protection, soil protection and health, water cycle and quality, air quality.

Local knowledge and culture can be considered as integral parts of agricultural biodiversity, because it is the human activity of agriculture which conserves this biodiversity. Indeed, most crop plants have lost their original seed dispersal mechanisms as a result of domestication, and so can no longer thrive without human input. Domestication started 10,000 years ago and has been followed by natural selection through exposure to different climates, pests, pathogens and weeds; by human selection for specific traits and market needs, as well as for socio-economic reasons; and by wide dispersal.

Crops and domestic animals are now found well beyond the limits of ecological tolerance of their immediate wild relatives, there is remarkable variability among and within crop landraces and animal breeds, and extraordinary ranges of adaptation. In the last 100 years, there has also been controlled plant and animal breeding by scientists which has allowed the recombination of diversity from widely different backgrounds, and the application of intense selection pressure. In terms of biological taxa, agricultural biodiversity includes the following:

— Higher plants: crops; harvested and managed wild plants for food; trees on farms, pasture and rangeland species

— Higher animals: domestic animals; wild animals hunted for food etc; wild and farmed fish

— Arthropods: mostly insects including pollinators (e.g., bees, butterflies), pests (e.g., grasshoppers, greenflies), and predators (e.g., wasps, beetles), and insects involved in the soil cycle (notably termites)

— Other macro-organisms: e.g., earthworms

— Micro-organisms: e.g., rhizobia, fungi, disease-producing pathogens

There are several distinctive features of agricultural biodiversity, compared to other components of biodiversity:

— agricultural biodiversity is actively managed by farmers;

— many components of agricultural biodiversity would not survive without this human interference;

indigenous knowledge and culture are integral parts of agricultural biodiversity management;

— many economically important farming systems are based on 'alien' crop species introduced from elsewhere. This creates a high degree of interdependence between countries for the genetic resources on which our food systems are based;

— as regards crop diversity, diversity within species is at least as important as diversity between species;

— because of the degree of human management, conservation of agricultural biodiversity in production systems is inherently linked to sustainable use—preservation through protected areas is of less relevance;

— nonetheless, in industrial-type agricultural systems, much crop diversity is now held exsitu in gene banks or breeders' materials rather than on-farm.

Elements of Agricultural Biodiversity

Crop Diversity

Of the 270,000 species of higher plants, about 7,000 species are used in agriculture, but only three (wheat, rice and maize) provide half of the world's plant-derived calorie intake. A substantial share of energy intake is also provided by meat, which is ultimately derived from forage and rangeland plants. Although world food production in the aggregate relies on few crop species, many more are important if production is disaggregated to regional, national or local levels. For example, in Central Africa cassava supplies over half of plant-derived energy intake, although at the global level the figure is only 1.6 per cent.

Beans, plantain, groundnut, pigeon pea, lentils, cowpeas and yams are the dietary staples of millions of poor people. Within individual countries, aggregate food supplies may be secured from few crop species but staples such as oca, teff, fonio, and bambarra nut can be vital in particular local pockets. A large number of other crops may be important as suppliers of other significant components of diet such as protein, fats, vitamins and minerals. But, outside the communities concerned, there is a lack of knowledge about the diversity and distribution of less utilised food and agriculture species.

Genetic diversity (variation within species) is vital for the evolution of agricultural species, and their adaptation to particular environments through a mixture of natural and human selection. In crop agriculture, for some species this selection has led to the development of many thousands of 'landraces' or 'farmers' varieties'.

Wild Plant Biodiversity

In addition to domesticated plants, wild species are important nutritionally and culturally to many people. Foods from wild species form an integral part of the daily diets of many poor rural households and are especially important during the hungry season or famines. They are an important source of vitamins, minerals and other nutrients, and also represent ready sources of income for cash-poor households. In Tanzania in 1988, for example, it was calculated that the value of all wild plant resources to rural communities was more than 8 per cent of agricultural GDP.

There are also wild relatives of crop plants which may supply useful genes through natural or artificial introgression. Neighbouring wild companion plants can harbour biocontrol agents useful in agriculture. Weed plants may be left to grow in fields in order to be harvested later for food. The term 'wild' may be misleading because it implies the absence of human influence. Many plant populations that have been considered wild are actually carefully nurtured by people, albeit less intensively than those cultivated in their fields. Thus, there is no obvious or strict divide between 'domesticated' and 'wild' food species; rather it is a continuum according to the extent of human intervention.

Below-ground Plant Biodiversity

Roots are responsible for nutrient and water uptake by crops. They physically stabilise soil structure against erosion and soil movement on steep slopes, and, in tropical systems, the contribution of roots to soil organic matter is proportionately larger than from above-ground inputs. The effects of roots on soil biophysical properties are particularly critical in impoverished farming systems where crop residues are at a premium for fuel and fodder.

Paradoxically, there has been little attention to the selection of rooting traits in cultivars by crop breeders, and much less research into the production, turnover and structure of rooting systems in tropical crops than into the above-ground components they support. In semi-arid regions, on soils of low inherent fertility, the phenology and distribution of roots determines water and nutrient availability for the crop during the growing season. Modern maize hybrids tend to show the rapid development of a large fine-root mass in topsoil which enables exploitation of superficial water and nutrients pools.

As drought conditions develop surface rooting systems are progressively exposed to moisture stress resulting in a progressive uncoupling of surface nutrient pools and available water at greater soil depths. Under stressed conditions many smallholder farmers prefer to plant traditional land races, which are adapted to different soil niches and associated environmental conditions. The genetic determinants and phenological plasticity of root architecture in these landraces has not been systematically investigated and hence provide options for selective breeding and management.

Microbial Biodiversity

Microbial biodiversity has been neglected over the years but is now a topic of global attention. This is due to the realisation that microbes contribute a wealth of gene pools that could be a source of material for transfer to plants to achieve traits such as stress tolerance and pest resistance, and large-scale production of plant metabolites. Of more immediate significance to farmers' production systems, microbes play varied roles in plant development and agriculture.

Microbial interactions with plant communities range from disease-producing pathogens to associations with plant rhizosphere, phyllosphere and spermosphere as free living entities or in well-associated symbiotic associations for nitrogen fixation or asmycorrhiza. Seed-borne microflora are instrumental in seed transmission of disease and thereby important in plant quarantine.

Arthropod Biodiversity

It is well known that insects, spiders and other arthropods often act as natural enemies of crop pests. But other components of arthropod diversity are also important in this respect. For example, research on Javanese rice fields has shown that arthropod communities are structured in such a way that the dynamics of seasonal succession consistently lead to high levels of pest suppression by natural enemies, with little chance of major pest outbreaks. Control of plant pests by natural enemies is often considered inadequate due to seasonal oscillations in populations: the pest population peaks before that of the natural enemies.

However, in the Javanese rice fields 'neutral' arthropods, mostly detritivores and planktonfeeders such as midges and mosquitoes, provide an alternative source of food for the natural enemies of rice plant pests, thus stabilising the populations of the natural enemies. In turn the detrivores are dependent on high levels of organic matter in the paddies which provides the food source for an array of micro-organisms (bacteria and phtytoplankton) and zooplantkton. This emphasises the importance of soil organic matter levels as a source of food for insects which offer an alternative food source for the natural enemies of plant insect pests, thereby stabilising natural enemy populations even in the absence of the plant pest and/or its host plant.

Arthropods are also important as pollinators of many crops. Bees and other pollinating insects are essential agents for the production of many crops, especially most major fruit and nut crops, many vegetable crops and a number of forage crops. Insect pollination is also required for seed production in crops such as soybean and sunflower. The estimated social worth of insect pollinators is of the order of several tens of billions of US dollars per annum.

Agricultural Biodiversity and Agroecosystem

Historically, the focus in agricultural biodiversity work has been on characterising and conserving species and genetic diversity. Now, however, there is increasing realisation of the importance of agricultural biodiversity at the ecosystems level, consistent with the 'ecosystem approach' as promoted by the Convention on Biological Diversity.

An ecosystem consists of a dynamic complex of plant, animal and micro-organism communities and their non-living environment interacting as a functional unit.

Thus agroecosystems need to be considered at several levels or scales, for instance, a leaf, a plant, a field/crop/herd/pond, a farming system, a land-use system or a watershed. These can be aggregated to form a hierarchy of agro-ecosystems. At a higher level still, the full assemblage of ecosystems constitutes the global biosphere. Ecological processes can also be identified at different levels and scales. Maintenance of agricultural biodiversity within the agro-ecosystem is necessary to ensure the continued supply of goods and services such as:

— evolution and crop improvement through plant breeding: it is now generally recognised that conservation off-site ('ex-situ') is only part of the action necessary to safeguard agricultural biodiversity for future evolution. The interaction between the environment, genetic resources and management practices that occurs 'in-situ' within agro-ecosystems ensures that a dynamic portfolio of agricultural biodiversity is maintained and adaptable to changing conditions;

— biological support to production: is provided by the organisms that make up the biological diversity of the agro-ecosystem. Earthworms and other soil fauna and microorganisms, together with the roots of plants and trees, ensure nutrient cycling; pests and diseases are kept in check by predators and disease control organisms, as well as by genetic resistances in crop plants themselves; and insect pollinators contribute to the cross-fertilisation of outcrossing crop plants. Agro-ecosystems vary in the extent that this biological support to production is replaced by external inputs:

— in industrial-type agricultural systems, they have been replaced to quite a significant extent by inorganic fertilisers and chemical pesticides; but in the many areas, agricultural biodiversity remains a significant provider;

— wider ecological functions: valuable ecological processes that result from the interactions between species and between species and the environment include the maintenance of soil fertility, water quality and climate regulation.

This is sometimes referred to as 'functional agricultural biodiversity', i.e. that which is necessary to sustain the ecological function of the agro-ecosystem, its structures and processes in support of food production and food security. Focussing attention on functional agricultural biodiversity can be a useful way of prioritising effort. There is no such thing as an a priori 'optimum' level of agricultural biodiversity for an agro-ecosystem; rather, the desirable level is determined by the prevailing local, natural and—equally importantly—socioeconomic circumstances.

The mix of agricultural biodiversity in any one agro-ecosystem is determined by a matrix of 'human' factors and feedback loops in addition to the underlying natural conditions. The factors determining levels of agricultural biodiversity in production systems are:

— underlying ecological conditions;

— farmers' skills in on-farm agricultural biodiversity management;

— farmers' access to useful agricultural biodiversity off-farm which is partly determined by connectedness, population pressure, communications, etc.;

— farmers' access to other capitals that can substitute for natural capital (for example, agrochemicals), which is significantly determined by prevailing explicit and implicit subsidies.

We can distinguish between more 'traditional' and more 'industrial' agricultural systems, although in reality most agricultural systems contain a unique and complex mixture of both traditional and industrial components. Agricultural systems that are more 'traditional' can be characterised as being less integrated into the market network, because of lack of financial capital, or lack of infrastructure such as roads and selling points, and lack of access to relevant agricultural research and extension. Hence, farmers in traditional-type systems place less reliance for their livelihoods on selling produce, and less reliance on buying external inputs for agricultural production such as chemical fertilisers and agro-chemicals.

Instead, they rely heavily on the available natural capital, in the form of quantity and quality of land, water resources, and agricultural biodiversity, to sustain their livelihoods. This includes, for example, the production of a wide range of food crops, fodder, medicine and building materials. Their emphasis may be on risk avoidance or minimisation, rather than on maximising production. This tends to produce a pattern of mixed farming in which a large number of species are cultivated, with considerable genetic diversity within species, and heavy use also made of wild plant diversity and non-plant agricultural biodiversity for both livelihoods and ecosystem functions and services.

In more traditional-type agricultural systems, farmers actively manage agricultural biodiversity on-farm in order to improve productivity and maintain sustainability. The key requirement is to enable farmers to continue to do this. In contrast, agricultural systems that are more 'industrial' are heavily integrated into the market system. Farmers produce largely for the market, and use the financial capital that is generated to fund investments in external inputs, as well as to provide other components of their livelihoods.

The ability to realise a financial surplus may be the result of having access to a combination of abundant and productive natural capital, infrastructure such as product and capital markets, and human capital such as education and access to information. Their superior access to capital assets means that farmers in industrial systems are often relatively less dependent on natural capital, and can focus on maximising production rather than minimising risk. This tends to result in a pattern of monoculture, focusing on a few profitable species and varieties and relying on off-farm conservation and breeding.

However, non-crop biodiversity may remain high in these systems, which may furthermore benefit significantly from functions and services provided by off-farm agricultural biodiversity (for example, watershed protection), and are of course reliant on crop diversity held off-farm for continued crop improvement. More industrial-type agricultural systems are only sustainable if the accompanying infrastructure is available to support them: plant breeding capacity, roads, markets, etc.

People often think in terms of a simple correlation between agricultural zone, agricultural system, and levels of agricultural biodiversity, ie. that more traditional-type agricultural systems are found in 'lower potential' agricultural zones and are reliant on high agricultural biodiversity, whilst more industrial-type agricultural systems are found in 'higher potential' agricultural zones and are characterised by low agricultural biodiversity.

Genetic Erosion and Conservation

Whether the introduction of new genetic material—be it

modern varieties, farmers' varieties, or by natural introgression—actually results in increasing or decreasing agricultural biodiversity on-farm will depend on:

— is the new germplasm added to the farmer's portfolio or does it replace existing germplasm?

— is the new germplasm uniform or heterogeneous?

— is it from a fixed or segregating population?

Also, the environmental, economic, and social dynamics outlined so far have different effects at field, farm, village, national, and global level. Some may have positive effects at one level but negative effects at another. Thus, the effects on agricultural biodiversity of these change processes are difficult to predict.

For example, in the case of crop diversity, decentralised plant breeding might, in some cases, lead to fewer varieties per farm. But if the result on each farm is different (because of specific adaptation during selection to the different environments and differences in farmer preference), then agricultural biodiversity may be maintained or enhanced when analysed at higher levels of aggregation. Furthermore, changes in agricultural biodiversity are difficult to measure at the genetic level.

The most common means of assessing erosion in farm level crop diversity is by counting named varieties, but this is different from actual genetic erosion because variety names do not necessarily correspond to cultivars/genetic content either geographically or over time. Better methods of assessment are required. Nonetheless, it appears that agricultural biodiversity is being eroded and the accompanying local knowledge of food producers is also under threat. In traditional-type agricultural systems, the main risk to agricultural biodiversity is from desertification, environmental degradation, and to some extent from species and varietal replacement.

In industrial-type agricultural systems, there is a high risk of genetic erosion on-farm through simplification of ecosystems, and species and varietal replacement. Everywhere, the genetic erosion of agricultural biodiversity is also exacerbated by the loss of forest cover, coastal wetlands and other 'wild' uncultivated areas. This leads to losses of wild relatives and losses of the wild foods that are essential for food provision. As regards plant diversity, up until the last decade or so, international scientists generally believed that the best way to conserve plant diversity was to collect samples from farmers' fields and preserve these in national and international 'gene banks'.

However, for various reasons it is now widely accepted that ex-situ conservation should be complemented by insitu or on-farm conservation, that is, conservation through sustainable use in farmers' fields. In-situ conservation is promoted in the Convention on Biological Diversity, and also forms a significant part of the Global Plan of Action for the Conservation and Sustainable Use of PGRFA. Proponents of this approach point out that the dynamism which makes on-farm conservation difficult to 'manage' conventionally is an essential part of the conservation process, whilst also recognising that on-farm conservation is a complement rather than a substitute for existing exsitu methods.

Benefits of Agricultural Biodiversity

Agricultural biodiversity provides many goods and services of environmental, economic and social importance and makes important contributions to sustainable livelihoods in a number of ways:

— agricultural biodiversity contributes directly to sustainable livelihoods in both traditional and industrial-type agricultural systems through production effects (crops, soil nutrient recycling, pest predators, etc);

— it also contributes to sustainable livelihoods in traditional and industrial-type agricultural systems through the provision of important ecosystem functions and services; and,

— it contributes to the livelihoods of a wide range of other stakeholders (public sector plant breeders and other agricultural research scientists, international biochemical companies, urban consumers in the North and South, the international gene bank system).

We need to take an integrated approach to analysing the contribution of agricultural biodiversity to sustainable livelihoods because of the significant spillover effects and feedback loops that operate. For example, areas of high agricultural biodiversity may provide environmental services needed to sustain monocultures in neighbouring industrial-type agricultural systems. As regards the feedback loops, remember that agricultural biodiversity assets are significantly affected by natural conditions and processes of evolution, as well as by production, consumption, conservation, and human components of the vulnerability context and transforming structures and processes.

Economists identify various categories of the values of agricultural biodiversity. However, the economic valuation of many aspects of agricultural biodiversity remains problematic. Direct uses of agricultural biodiversity include a range of products which provide dietary diversity and make important nutritional contributions (provision of minerals, vitamins and protein; hunger crops). They include:

— Consumptive uses: goods that do not appear in national economic statistics, but which local people need (e.g. medicinal plants, wild vegetables, building materials), can be valued at the cost of market alternatives.

— Productive uses (goods sold in commercial markets) are conventionally valued at the net price at the point of sale.

Additionally crop diversity can generate improvements in yields through plant breeding. For example, genetic improvements in US crops were responsible for increasing the value of the harvest by an average of $1 billion per year from 1930 to 1980. Indirect uses of agricultural biodiversity include production effects such as adaptation to lower input conditions; specific adaptation (intra-farm and inter-farm); reduction of risk; potential for high biological production; and having a range of varieties and species with complementary agro- ecological requirements.

Indirect uses also include ecosystem services: biodiverse agriculture provides more of these important services than does monoculture. Some ecosystem services can be valued relatively straightforwardly, for example wild insects pollinating crops can be valued at the incremental value of the crop, or the cost of hiring honey bees. Others, such as CO_2 absorption by plant communities, are much harder. However, the value of these services is rarely captured in a market. Indeed, the value of ecosystem services is inadequately captured using conventional economic analysis.

The 'option' value of biodiversity is the potential of agricultural biodiversity to provide economic benefit to human society in the future. Two components of this are:

— *Insurance value*: insurance against future adverse conditions, as needs are constantly changing and because genetic resources may later prove to provide useful characteristics, for example resistance to new diseases or adaptability to changed climatic conditions; and

— *Exploration value*: agricultural biodiversity represents a treasure chest of potentially valuable but as yet unknown resources.

The values of agricultural biodiversity are often not completely captured by the relevant stakeholders, which has important implications for agricultural biodiversity conservation and use:

— some values of agricultural biodiversity are realised at higher levels of aggregation—for example, the value for reducing variability in food yields and thus prices world-wide is felt on global markets, not by individual farmers. This limits the incentives provided to individual farmers or national governments to invest in agricultural biodiversity conservation;

— values such as 'exploration value' and 'option value' are usually public goods under current institutional arrangements. This means that, although they may provide benefits to certain groups, these benefits are not fully appropriable under existing property rights systems. The exception to this being where exclusive bio-prospecting rights are granted, but this usually applies to medicinal plants rather than agricultural crops;

— much of the value of agricultural biodiversity is not divisible. This means that, although different groups of stakeholders have contributed to agricultural biodiversity management and development, and should therefore each receive a share of the benefit, in practice some groups are better able to appropriate a large share of the benefit for themselves.

To date, stakeholders who use agricultural biodiversity directly, such as farmers in both traditional and industrial-type agriculture, and providers and users of traditional medicine, have been weaker in terms of voice and market power compared to those with an interest in controlling access to it. The emergence of a strong private sector in the shape of international agro-chemical companies has greatly complicated matters. The private sector demands effective intellectual property protection before investing in technology development, so the technologies and knowledge produced by the private sector are no longer public goods.

On the other hand some of the goods and services provided by agricultural biodiversity, such as most ecosystem functions, will always remain public goods; and incentives or public support may be required to ensure the continued provision of these positive externalities. Overall, there have been powerful forces pushing for a reduction in agricultural biodiversity on-farm, through the promotion of chemical fertilisers, uniform crops and varieties, etc. This tendency may be strengthened by increasing consolidation in the 'life science' industry and the acquisition of seed companies by chemical companies. On the other hand, there are also some countervailing changes:

— the end to the global system of agricultural subsidies promoting industrial-type agriculture (through the current WTO negotiations, etc) is leading to the development of new agricultural practices and technologies that may be more biodiversity-friendly;

— the increasing voice of consumers demanding ecologically-friendly agricultural production processes;

— the increasing voice of farmers' and civil society organisations;

— the increasing recognition of cultural values and indigenous technical knowledge in important international treaties such as the Convention on Biological Diversity;

— the increase in the number of international treaties and agreements promoting conservation, sustainable use and benefit-sharing in agricultural biodiversity.

This analysis of agricultural biodiversity points to the inevitability of conflicts between the vastly differing interests of diverse stakeholders in agricultural biodiversity, exacerbated by their dramatically different degrees of effective voice and market power. This points towards the importance of treaties for reaching agreement on agricultural biodiversity issues, as well as institutions for effective management, at the local, national and international levels. There is a need to promote mechanisms that return a fair proportion of the benefits to the stakeholders who manage agricultural biodiversity at the local level. Additionally, non-market uses of agricultural biodiversity (for example, the provision of ecosystem services and functions) require public support to ensure their continued provision.

Role of Agricultural Biodiversity in Poverty

There is some evidence that agricultural biodiversity—particularly plant diversity—is concentrated in areas of poverty: in general, there is more plant diversity in developing countries than in developed countries; further, plant diversity tends be concentrated in the poorest, least developed regions of countries. This has led to a view that development and agricultural biodiversity are in opposition, and that economic development should involve the conversion' of diverse areas to 'more productive' areas. But this is an over-simplification of the issue in at least three respects:

- it is difficult to compare agricultural biodiversity across zones, because levels of diversity are different for different sub-sets of agricultural biodiversity in different agro-ecological zones, eg, in intensive rice systems of South and South-East Asia, crop diversity is relatively low but non-crop biodiversity can be high;
- agricultural biodiversity is essential for rich countries and industrial-type agriculture for continued evolution and agricultural improvement, although it is no longer usually maintained on-farm in these areas;
- the relationship between agricultural biodiversity and poverty at the micro level is not clear cut, as is explained in the following paragraphs.

Closer examination of many 'less developed' areas with biodiverse agricultural systems shows that farmers often choose to maintain local crop germplasm because in these areas, and under current economic conditions, it spreads the risk better than the alternatives from formal sector plant breeding. However, there are important variations in different households' dependence on agricultural biodiversity within communities. On the one hand, researchers have found that within any given community, crop diversity is often handled more by richer farmers. On the other hand, there are also clear cases where poor or vulnerable groups are highly dependent on other aspects of agricultural biodiversity (minor crops, wild plants, soil biota, insects) and may maintain it more carefully. These groups are often directly dependent on agricultural biodiversity for both on- and offfarm livelihood activities.

Loss of this biodiversity can be associated with heavy livelihood losses through undermining their production choices, food security, and increasing their exposure to risk. Thus, the correlation between agricultural biodiversity and poverty does not indicate a causal relationship, only that a location-specific approach to development is required. Ultimately, the most appropriate blend of agricultural biodiversity in farmers' portfolios depends on the precise local context. Hence, there is a need to work with local farmers and communities in a participatory manner to identify opportunities for action and the most appropriate means of implementation.

Undoubtedly, agricultural biodiversity is decreasing world-wide due to the combined effects of 'specialisation, harmonisation and homogenisation'—all components of globalisation. Conversion for economic gain is a fact in both traditional and industrial-type agricultural systems. People still convert in traditional-type agricultural systems to lower agricultural biodiversity livelihood strategies due, amongst other causes, to prevailing economic distortions which are institutionalised in the current global economic system. These include, for example, input subsidies, agricultural extension messages, or widespread distribution of modern seeds in emergency relief packages.

For politically, economically and socially marginalised groups, protection, maintenance of, and improved access to, agricultural biodiversity can often contribute more to sustainable livelihoods than conversion, as their traditional entitlements to agricultural biodiversity may be stronger than their market access to agricultural production inputs. Note however that there are often good reasons for adding some new genetic material to farmers' variety portfolios, to fill particular niches (storability, taste, etc) or to cope with change in the agro-ecosystem (e.g. climate change requiring shorter duration materials). It is therefore highly inappropriate to promote large-scale abandonment of biodiverse agriculture. But the challenge is to create a new enabling environment that makes returns to the maintenance of agricultural biodiversity more sustainable and more accurately reflect agricultural biodiversity's true value to the livelihoods of different stakeholders.

A new approach to agricultural biodiversity is needed, in which local, national and global action contribute to the management of agricultural biodiversity. This involves:

— cross-sectoral action (agricultural biodiversity issues have relevance beyond natural resources management);

— a combination of policy and area-based approaches;

— a decentralised knowledge-intensive approach to technology development where farmers are full participants in the process;

— strengthening of local institutions;

— a high degree of policy input into arrangements for managing and sharing agricultural biodiversity because of the significant differences in agricultural biodiversity between people, countries, and regions.

It is essential to change the economic incentives, and institutional and policy barriers that currently discourage the sustainable use of agricultural biodiversity. It is also essential to correct the current policy, research and implementation activities which subsidise the replacement of goods and services provided by agricultural biodiversity by external inputs, and thereby create an inertia reinforcing the industrial-type agriculture model.

Agricultural biodiversity is often viewed as a resource stock that can be drawn down in order to contribute to strengthening people's livelihoods, but in reality it is more of a dynamic system than a stock. Agricultural biodiversity has social and economic as well as environmental and biological components, and is subject to human as well as natural selection pressures. Therefore, although it is often regarded as part of 'natural capital', in fact it also has important components of 'human capital'.

Care should be taken not to undermine these traditional approaches nor to sweep away unnecessarily the agricultural biodiversity on which they are based. But a purely conservationist approach must also be avoided. Farmers in traditional systems, just as much as those in industrial farming systems, need access to modern varieties as well as to farmers' varieties, and to the whole range of genetic material in between. At the local level, it is important to remember that different agro-ecosystems may require different approaches.

In traditional-type agriculture, the aim is to maximise the contribution of agricultural biodiversity to sustainable rural livelihoods. This may involve strengthening farmer/community management of existing agricultural biodiversity on-farm and increasing access to a range of agricultural biodiversity and to related skills and technologies in order to use agricultural biodiversity more effectively. In industrial-type agriculture, the emphasis may be more on modifying the genetic basis of plant breeding through base broadening and participatory approaches, and strengthening the ex-situ conservation of agricultural biodiversity. These are actions that require national level action. Even in industrial-type agriculture, some vulnerable groups may be dependent on on-farm agricultural biodiversity, and in such areas it will be important to protect this resource base at the local level.

In both areas, there are good reasons to promote integrated pest management which relies on the biodiversity of natural enemies of pests and good management of soil biota. At the national level, it is important to provide an enabling environment that will support local level actions aimed at strengthening the livelihoods of the rural poor. But there is also another dimension to consider at this level. The well-being of the urban-poor and the non-food producing rural poor, is improved by a plentiful and cheap food supply.

Thus a total poverty elimination strategy will require appropriate institutional arrangements (including those for conservation and access of genetic resources, and plant breeding) at the national level to support sustainable crop production in both traditional and industrial-type agricultural systems. Agricultural biodiversity underpins food security at the global level too. Thus, in addition to supporting local and national level needs, global level policies and programmes should also ensure adequate conservation of agricultural biodiversity and sharing of its benefits in the aggregate.

Poverty Reduction Policies

Opportunities at the global level for supporting the use of agricultural biodiversity for poverty reduction fall into two categories: firstly, those that promote the activities described above at national and local levels; and, secondly, some actions that can most effectively be carried out only at supra-national level. These include long-term conservation; genetic enhancement of major crop gene pools; international frameworks for conservation, sustainable use and benefit sharing. These actions can be implemented through three channels, as follows:

The International Agricultural Research System

Using agricultural biodiversity for sustainable livelihoods requires a continuing commitment to research, in order to improve our understanding of various as yet unclear ecological and economic relationships. The key research gaps in ecology, economics and agriculture include:

— Environment research on:

a) the contribution agricultural biodiversity can make to sustainable agriculture;

b) defining agreed indicators for agricultural biodiversity assessments;

c) factors determining the rate of extinction in agricultural biodiversity;

d) co-dependency between different components of agricultural biodiversity.

— Socio-economic research on the valuation of ecosystem functions and services.

— Plant sciences research on the role of roots in crop growth and in agro-ecosystem functioning.

The International Agricultural Research Centres of the CGIAR, as the backbone of the international agricultural research system, should focus their work related to agricultural biodiversity on:

— supporting the international network of ex-situ gene banks;

— facilitating genetic enhancement or base broadening of major crop gene pools available to national public and private plant breeders;

— supporting a decentralised approach to plant breeding;

— monitoring the progress and outcomes of gene transfer carefully (note that at the October 1998 Centres Week, the CG system decided that none of its plant breeders will use 'any genetic system designed to prevent germination');

— contributing to the capacity building of national programmes, with wide stakeholder involvement, in order that national programmes as well as farmers and their communities can exert a 'demand pull' on the international agricultural research system;

— developing methodologies in support of local agricultural biodiversity conservation and sustainable use.

Aid, Development and Environment Programmes

Agricultural biodiversity should be included in development education programmes. And awareness-raising amongst Northern consumers is also important in order to publicise the ecological and social 'foot prints' of their food consumption patterns and to promote the consumption of food from biodiversity-friendly agricultural production systems. In this regard, it may be possible to build on the current stimulation of markets for organic produce, and the promotion of ethically traded goods.

International Agreements and Decision-making Fora

The framework of international law and regulations should:

— facilitate access to genetic resources and related information and technologies to prevent monopolistic conditions. These are influenced by IPR law and exemptions, specific regulations on access, and potentially by anti-trust measures;

— provide for biosafety, in terms both of protecting the environment and of avoiding damage to livelihoods through vulnerability to agricultural systems;

— provide for the various aspects of Farmers Rights required for farmers and communities to conserve, develop and share in the benefits arising from the use of agricultural biodiversity.

Concerning international agreements under development or review the most critical are:

— The International Undertaking on Plant Genetic Resources is currently under revision and needs to include agreements to facilitate access to plant genetic resources, with minimal restrictions, and for the full implementation of Farmers' Rights including the right to resow saved seed. Funding issues, possibly including funding of the Global Plan of Action, are also under discussion, as are several critical issues of Farmers Rights;

— Various matters being discussed under the Convention on Biological Diversity, including matters relating to access and benefit sharing, traditional knowledge, and implementation of the Cartagena Protocol on biosafety, as well as implementation of thematic work programmes particularly the programme of work on agricultural biodiversity.

— The upcoming review of the agreement on Trade Related Intellectual Property Rights should maintain the option to exclude plants and animals from patenting (while providing a sui generis system for plant varieties). International obligations such as the WTO/TRIPs agreement should be implemented in a way consistent with national interests.

6

Biodiversity in India

India is one of the world's richest countries in terms of its vast array of biological diversity. The Western Ghats and North East India is recognised as an international biodiversity hotspot. It has been estimated that over 81,000 species of fauna and 47,000 species of flora are found in this country so far. Of the estimated 47,000 plant species, about 15,000 flowering species are endemic to India. The relationship between local communities and the biological diversity in India is a very intricate one, whereby two-thirds of our population are heavily dependent upon the biodiversity for their survival.

The destruction of biodiversity signifies the destruction of people's livelihoods and survival. At present, around the world, at least three species a day are being depleted, or 1,200 species every 400 days. At the current rates of extinction, a quarter of the world's known and identified species (over 400,000) will have vanished by the end of this century—thats only three years away. Among the larger animals in India, 79 species of mammals, 44 of birds, 15 of reptiles, and three of amphibians are threatened.

Nearly 1500 plant species are considered endangered. Flowering plants and vertebrate animals have recently become extinct at a rate estimated at 50 to 100 times the average expected natural rate. The loss of even one species causes changes that are complex, unpredictable and incomprehensible. According to the FAO, replacement of local varieties forms the largest contribution to biodiversity erosion. Replacement of local varieties by a select number of high yielding hybrid varieties (mono-cropping) has been vigorously promoted under the Green Revolution during the 1960s in order to increase production of food crops.

In India, monocropping has resulted in the vast depletion of indigenous species of flora and fauna. Despite being challenged by eminent scientists around the world, the Green Revolution's tenets continue to influence agriculture in the developing countries today. Most Indians live in a biodiversity based economy. Therefore, biodiversity and indigenous knowledge are central to the economic security and subsequently, the national security of our country. Biopiracy refers to the process by which the rights of indigenous cultures to these resources and knowledge are erased and replaced by monopoly rights for those who have exploited indigenous knowledge and biodiversity.

Biopiracy is occurring through the claiming of IPRs by TNCs, who are making billions of dollars by usurping our traditional knowledge systems and genetic resources from the domain of the commons. IPRs regimes only recognise and provide protection to formal innovators, not to indigenous informal innovators. Therefore, traditional knowledge evolved and utilised by informal innovators is being pirated by formal innovators, who perform mere translations and minor modifications and then seek patents, claiming the knowledge as well as life forms as their private property.

Most biodiversity related intellectual property rights are either the enclosure of non-western traditions of science (for example; neem, turmeric, and phyllanthus niruri) or the enclosure of the intellectual commons of contemporary western traditions of science. Biopiracy through IPRs has arisen as a result of the devaluation and invisibility of indigenous knowledge systems and the lack of existing protection of these systems. The protection of indigenous knowledge systems as systems of innovation and the prevention of piracy of biodiversity requires a widening of legal regimes beyond the existing IPR regimes such as patents.

WETLANDS

India has a rich variety of wetland habitats. The total area of wetlands in India is 58,286,000ha, or 18.4% of the country, 70% of which comprises areas under paddy cultivation. A total of 1,193 wetlands, covering an area of about 3,904,543 ha, were recorded in a preliminary inventory coordinated by the Department of Science and Technology, of which 572 were natural. Two sites—Chilka Lake (Orissa) and Keoladeo National Park (Bharatpur)—have been designated under the Convention of Wetlands of International Importance (Ramsar Convention) as being especially significant waterfowl habitats.

The country's wetlands are generally differentiated by region into eight categories:

— the reservoirs of the Deccan Plateau in the south, together with the lagoons and the other wetlands of the southern west coast;

— the vast saline expanses of Rajasthan, Gujarat and the gulf of Kachchh; freshwater lakes and reservoirs from Gujarat eastwards through Rajasthan and Madhya Pradesh;

— the delta wetlands and lagoons of India's east coast (Chilka Lake); the freshwater marshes of the Gangetic Plain;

— the floodplain of the Brahmaputra; the marshes and swamps in the hills of north-east India and the Himalayan foothills;

— the lakes and rivers of the montane region of Kashmir and Ladakh; and

— the mangroves and other wetlands of the island arcs of the Andamans and Nicobars.

FORESTS

India possesses a distinct identity, not only because of its geography, history and culture but also because of the great diversity of its natural ecosystems. The panorama of Indian forests ranges from evergreen tropical rain forests in the Andaman and Nicobar Islands, the Western Ghats, and the north-eastern states, to dry alpine scrub high in the Himalaya to the north. Between the two extremes, the country has semi-evergreen rain forests, deciduous monsoon forests, thorn forests, subtropical pine forests in the lower montane zone and temperate montane forests.

One of the most important tropical forests classifications was developed for Greater India and later republished for present-day India. This approach has proved to have wide application outside India. In it 16 major forests types are recognised, subdivided into 221 minor types. Structure, physiognomy and floristics are all used as characters to define the types. The main areas of tropical forest are found in the Andaman and Nicobar Islands; the Western Ghats, which fringe the Arabian Sea coastline of peninsular India; and the greater Assam region in the north-east.

Small remnants of rain forest are found in Orissa state. Semi-evergreen rain forest is more extensive than the evergreen formation partly because evergreen forests tend to degrade to semi-evergreen with human interference. There are substantial differences in both the flora and fauna between the three major rain forest regions. The Western Ghats Monsoon forests occur both on the western (coastal) margins of the ghats and on the eastern side where there is less rainfall. The forests contain several tree species of great commercial significance, but they have now been cleared from many areas.

In the rain forests there is an enormous number of tree species. At least 60 percent of the trees of the upper canopy are of species which individually contribute not more than one percent of the total number. Clumps of bamboo occur along streams or in poorly drained hollows throughout the evergreen and semi-evergreen forests of south-west India, probably in areas once cleared for shifting agriculture. The tropical vegetation of north-east India typically occurs at elevations up to 900 m. It embraces evergreen and semi-evergreen rain forests, moist deciduous monsoon forests, riparian forests, swamps and grasslands.

Evergreen rain forests are found in the Assam Valley, the foothills of the eastern Himalayas and the lower parts of the Naga Hills, Meghalaya, Mizoram, and Manipur where the rain fall exceeds 2300 mm per annum. In the Assam Valley the giant Dipterocarpus macrocarpus and Shorea assamica occur singly, occasionally attaining a girth of up to 7 m and a height of up to 50 m. The monsoon forests are mainly moist sal Shorea robusta forests, which occur widely in this region. The Andamans and Nicobar islands have tropical evergreen rain forests and tropical semi-evergreen rainforests as well as tropical monsoon moist monsoon forests.

The tropical evergreen rain forest is only slightly less grand in stature and rich in species than on the mainland. The dominant species is Dipterocarpus grandiflorus in hilly areas, while Dipterocarpus kerrii is dominant on some islands in the southern parts of the archipelago. The monsoon forests of the Andamans are dominated by Pterocarpus dalbergioides and Terminalia spp.

Marine Environment

The nearshore coastal waters of India are extremely rich fishing grounds. The total commercial marine catch for India has stabilised over the last ten years at between 1.4 and 1.6 million tonnes, with fishes from the clupeoid group accounting for approximately 30% of all landings. In 1981 it was estimated that there were approximately 180,000 non-mechanised boats carrying out small-scale, subsistence fishing activities in these waters. At the same time there were about 20,000 mechanised boats and 75 deep-sea fishing vessels operating mainly out of ports in the states of Maharashtra, Kerala, Gujarat, Tamil Nadu and Karnataka.

Coral reefs occur along only a few sections of the mainland, principally the Gulf of Kutch, off the southern mainland coast, and around a number of islands opposite Sri Lanka. This general absence is due largely to the presence of major river systems and the sedimentary regime on the continental shelf. Elsewhere, corals are also found in Andaman, Nicobar, and Lakshadweep island groups although their diversity is reported to be lower than in south-east India. Indian coral reefs have a wide range of resources which are of commercial value.

Exploitation of corals, coral debris and coral sands is widespread on the Gulf of Mannar and Gulf of Kutch reefs, while ornamental shells, chanks and pearl oysters are the basis of an important reef industry in the south of India. Sea fans and seaweeds are exported for decorative purposes, and there is a spiny lobster fishing industry along the south-east coast, notably at Tuticorin, Madras and Mandapam Commercial exploitation of aquarium fishes from Indian coral reefs has gained importance only recently and as yet no organised effort has been made to exploit these resources. Reef fisheries are generally at the subsistence level and yields are unrecorded.

Other notable marine areas are seagrass beds, which although not directly exploited are valuable as habitats for commercially harvested species, particularly prawns, and mangrove stands. In the Gulf of Mannar the green tiger prawn Penaeus semisulcatus is extensively harvested for the export market. Seagrass beds are also important feeding areas for the dugong Dugong dugon, plus several species of marine turtle. Five species of marine turtle occur in Indian waters: Green turtle Chelonia mydas, Loggerhead Caretta caretta, Olive Ridley Lepidochelys olivacea, Hawksbill Eretmochelys imbricata and Leatherback Dermochelys coriacea.

Most of the marine turtle populations found in the Indian region are in decline. The principal reason for the decrease in numbers is deliberate human predation. Turtles are netted and speared along the entire Indian coast. In south-east India the annual catch is estimated at 4,000-5,000 animals, with C. mydas accounting for about 70% of the harvest. C. caretta and L. olivacea are the most widely consumed species. E. imbricata is occasionally eaten but it has caused deaths and so is usually caught for its shell alone. D. coriacea is boiled for its oil which is used for caulking boats and as protection from marine borers. Incidental netting is widespread. In the Gulf of Mannar turtles are still reasonably common near seagrass beds where shrimp trawlers operate, but off the coast of Bengal the growing number of mechanised fishing boats has had the effect of increasing incidental catch rates.

Species Diversity

India contains a great wealth of biological diversity in its forests, its wetlands and in its marine areas. This richness is shown in absolute numbers of species and the proportion they represent of the world total. India has a great many scientific institutes and university departments interested in various aspects of biodiversity. A large number of scientists and technicians have been engaged in inventory, research, and monitoring. The general state of knowledge about the distribution and richness of the country's biological resources is therefore fairly good.

Inventories of birds, mammals, trees, fish and reptiles are moderately complete. Knowledge of special interest groups such as primates, pheasants, bovids, endemic birds, orchids, and so on, is steadily improving through collaboration of domestic scientists with those from overseas. The importance of these biological resources cannot be overestimated for the continued welfare of India's population.

Endemic Species

India has many endemic plant and vertebrate species. Among plants, species endemism is estimated at 33% with c. 140 endemic genera but no endemic families. Areas rich in endemism are north-east India, the Western Ghats and the north-western and eastern Himalayas. A small pocket of local endemism also occurs in the Eastern Ghats. The Gangetic plains are generally poor in endemics, while the Andaman and Nicobar Islands contribute at least 220 species to the endemic flora of India.

WCMC's Threatened Plants Unit (TPU) is in the preliminary stages of cataloguing the world's centres of plant diversity; approximately 150 botanical sites worldwide are so far recognised as important for conservation action, but others are constantly being identified. Five locations have so far been issued for India: the Agastyamalai Hills, Silent Valley and New Amarambalam Reserve and Periyar National Park, and the Eastern and Western Himalaya. Endemism among mammals and birds is relatively low. Only 44 species of Indian mammal have a range that is confined entirely to within Indian territorial limits.

Four endemic species of conservation significance occur in the Western Ghats. They are the Lion-tailed macaque Macaca silenus, Nilgiri leaf monkey Trachypithecus johni, Brown palm civet Paradoxurus jerdoni and Nilgiri tahr Hemitragus hylocrius. Only 55 bird species are endemic to India, with distributions concentrated in areas of high rainfall. They are located mainly in eastern India along the mountain chains where the monsoon shadow occurs, south-west India, and the Nicobar and Andaman Islands.

In contrast, endemism in the Indian reptilian and amphibian fauna is high. There are around 187 endemic reptiles, and 110 endemic amphibian species. Eight amphibian genera are not found outside India. They include, among the caecilians, Indotyphlus, Gegeneophis and Uraeotyphlus; and among the anurans, the toad Bufoides, the microhylid Melanobatrachus, and the frogs Ranixalus, Nannobatrachus and Nyctibatrachus. Perhaps most notable among the endemic amphibian genera is the monotypic Melanobatrachus which has a single species known only from a few specimens collected in the Anaimalai Hills in the 1870s. It is possibly most closely related to two relict genera found in the mountains of eastern Tanzania.

Threatened Species

India contains 172 species of animal considered globally threatened by IUCN, or 2.9% of the world's total number of threatened species. These include 53 species of mammal, 69 birds, 23 reptiles and 3 amphibians. India contains globally important populations of some of Asia's rarest animals, such as the Bengal Fox, Asiatic Cheetah, Marbled Cat, Asiatic Lion, Indian Elephant, Asiatic Wild Ass, Indian Rhinoceros, Markhor, Gaur, Wild Asiatic Water Buffalo etc.

A workshop held in 1982 indicated that as many as 3,000-4,000 higher plants may be under a degree of threat in India. Since then, the Project on Study, Survey and Conservation of Endangered species of Flora (POSSCEP) has partially documented these plants. The protection of wildlife has a long tradition in Indian history. Wise use of natural resources was a prerequisite for many hunter-gatherer societies which date back to at least 6000 BC. Extensive clearance of forests accompanied the advance of agricultural and pastoral societies in subsequent millennia, but an awareness of the need for ecological prudence emerged and many so-called pagan nature conservation practices were retained.

As more and more land became settled or cultivated, so these hunting reserves increasingly became refuges for wildlife. Many of these reserves were subsequently declared as national parks or sanctuaries, mostly after Independence in 1947. Examples include Gir in Gujarat, Dachigam in Jammu & Kashmir, Bandipur in Karnataka, Eravikulum in Kerala, Madhav in Madhya Pradesh, Simlipal in Orissa, and Keoladeo, Ranthambore and Sariska in Rajasthan. Wildlife, together with forestry, has traditionally been managed under a single administrative organisation within the forest departments of each state or union territory, with the role of central government being mainly advisory.

There have been two recent developments. First, the Wildlife (Protection) Act has provided for the creation of posts of chief wildlife wardens and wildlife wardens in the states to exercise statutory powers under the Act. Under this Act, it is also mandatory for the states to set up state wildlife advisory boards. Secondly the inclusion of protection of wild animals and birds in the concurrent list of the constitution, has proved the union with some legislative control over the states in the conservation of wildlife.

The situation has since improved, all states and union territories with national parks or sanctuaries having set up wildlife wings. The adoption of a National Policy for Wildlife Conservation in 1970 and the enactment of the Wildlife (Protection) Act in 1972 lead to a significant growth in the protected areas network, from 5 national parks and 60 sanctuaries to 69 and 410 respectively, in 1990. The network was further strengthened by a number of national conservation projects, notably Project Tiger, initiated in April 1973 by the Government of India with support from WWF, and the crocodile Breeding and Management Project, launched on 1 April, 1975 with technical assistance from UNDP/FAO.

Biodiversity on Western Ghats

The Western Ghats are a chain of highlands running along the western edge of the Indian subcontinent, from Bombay south to the southern tip of the peninsula, through the states of Maharashtra, Karnataka, Kerala and Tamil Nadu. Covering an estimated area of 159,000 sq. km, the Western Ghats are an area of exceptional biological diversity and conservation interest, and are one of the major Tropical Evergreen Forest regions in India. As the zone has already lost a large part of its original forest cover it must rank as a region of great conservation concern.

The small remaining extent of natural forest, coupled with exceptional biological richness and ever increasing levels of threat, are factors which necessitate major conservation inputs. There are currently seven national parks in the Western Ghats with a total area of 2,073 sq. km and 39 wildlife sanctuaries covering an area of about 13,862 sq. km (8.1%). The management status of the wildlife sanctuaries in this part of India varies enormously. Tamil Nadu's Nilgiri wildlife sanctuary, for example, has no human inhabitants, small abandoned plantation areas and no produce exploitation, while the Parambikulam wildlife sanctuary in Kerala includes considerable areas of commercial plantations and privately owned estates with heavy resource exploitation.

International Programmes and Conventions

India participates with many international agreements and programmes concerned with aspects of nature conservation and sustainable development. These range from legal instruments such as the Convention on Biological Diversity, which place obligations on those nations which become contracting parties, to scientific programmes such as the UNESCO Man and the Biosphere Programme, a global programme of international scientific cooperation. Examples of agreements and programmes with which India is collaborating include:

Convention on International Trade in Endangered Species (CITES)

Since India became a party to CITES on 18th October 1976 it has provided data annually to the CITES secretariat on the trade of endangered species through its CITES Management Authority. The text of the CITES convention along with the CITES appendices are provided.

World Heritage Convention

India ratified the World Heritage Convention in 1977 and since then five natural sites have been inscribed as areas of 'outstanding universal value'. These sites are:

— Kaziranga National Park

— Keoladeo National Park

— Manas National Park

— Sundarbans National Park

— Nanda Devi National Park

Convention on Biological Diversity

India signed the Convention on Biological Diversity on 5th June 1992, ratified it on 18th February 1994 and brought it into force on 19th May 1994. This convention will provide a framework for the sustainable management and conservation of India's natural resources.

Ramsar (Wetlands) Convention

India has been a contracted party to the Ramsar Convention since 1st February 1982. India has now six sites covering some 192,973 hectares of important wetlands.

7

Causes and Consequences of Biodiversity Loss

Causes of Biodiversity Loss

The Global Biodiversity Strategy identifies six fundamental causes of biodiversity loss:

— unsustainably high rates of natural resource consumption and human population growth;

— steadily narrowing spectrum of traded products from agriculture and forestry, and introduction of exotic species associated with agriculture, forestry and fisheries (including bio-engineered species);

— economic systems and policies that fail to value the environment and its resources;

— inequity in ownership and access to natural resources, including the benefits from use and conservation of biodiversity;

— inadequate knowledge and inefficient use of information;

— legal and institutional systems that fail to protect against unsustainable exploitation.

The above are a mixture of both direct and indirect causes and are described below.

Direct Causes

Habitat loss and degradation

Habitat loss and degradation is by far the leading cause of biodiversity loss in the region. In some cases, habitat alteration is intensive and large-scale, for example, when native forest is converted to plantation, or when a large dam inundates an area. In other instances, habitats are gradually eroded over time, as when a native forest or grassland is fragmented by expanding agricultural practice and/or population pressures. In either case, plant species are lost and many animal species are forced from their natural habitats. It is estimated that at least three quarters of all threatened bird species are in danger because of the pressures on their natural habitats, and that habitat loss is also the principal factor for the decline of at least three-quarters of all mammal species.

In areas where forest degradation and conversion have been most intense, such as South and Southeast Asia, a significant proportion of the endemic primate species face extinction. Species of reptiles and amphibians are also declining for similar reasons; habitat loss accounts for the decline of 68 per cent of all threatened reptile species and 58 per cent of threatened amphibian species.

Although degradation of terrestrial habitats such as forests may gain the most attention, both marine and freshwater habitats are also under serious pressure from human activities such as urbanisation and agricultural expansion. For example, river engineering works such as the construction of dams and/or levees have the effect of either inundating or drying up wetlands and backwaters which are important fish spawning grounds, and changing flow regimes to downstream lakes and estuaries, thereby altering their ecologies.

A substantial proportion of marine ecosystems are also at risk, primarily due to the direct activities associated with coastal development, including dredging, filling, breakwater construction, mining and drilling, but also from secondary effects such as pollution and increased marine traffic. The impacts of habitat pollution on biodiversity can be instantaneous or cumulative. For instance, oil wastes can asphyxiate and/or poison a wide range of marine life—from algae to seabirds—whereas other contaminants, such as radioactive wastes, pesticides, and other toxic chemicals, can take a while to build up and cause harm within individual organisms, especially within species high on the food chain.

In addition, the impact of pollution can be secondary, for example, certain species of algae will capitalise on high nutrient conditions introduced by some forms of pollution, undergoing massive population explosions which can in turn introduce toxins which are harmful to marine life such as fish and shellfish, or to consumers of those produce, or which can lower water clarity and oxygen content to the extent that marine life is depleted. Coral bleaching is another frequent outcome of pollution-induced stress.

Excessive use of biological resources

Throughout South and East Asia, a major factor in the excessive exploitation of wildlife has been over-hunting in response to the unsustainable demands of both national and international trade in goods such as animal parts for traditional medicine or curios. Species which have been particularly affected include the tiger, which has been hunted almost to extinction in South Asia, and the seahorse, which is harvested in numbers approaching 20 million per year—a rate which is unlikely to be sustainable because of their low reproductive rate, complex social behaviour and low mobility. Already, it is thought that some 36-seahorse species are threatened by this growing, unregulated harvest.

Tortoise and river turtle species also are exploited intensively in certain areas of the region; in Southeast Asia they have long been an important source of meat and eggs, but there is also now a burgeoning international trade for their use in traditional medicine, in countries such as People's Republic of China. At least five turtle species involved in this trade are now candidates for the most stringent listing available under the Convention on International Trade in Endangered Species of Wild Flora and Fauna (CITES). Uncontrolled harvesting of fish is playing a major role in jeopardizing freshwater and marine ecosystems and their native biota, resulting in a situation whereby about 68 per cent of all threatened marine fish species in the region suffer from over-exploitation.

A prime example is the sturgeon in Central Asia, where poaching is now so widespread that the few remaining stocks of this fish are nearly gone. Aside from overfishing of commercial stocks, poor management practices are also to blame for the decline in marine resources, which include blast fishing, fishing with cyanide and other poisonous chemicals, muro-ami netting and coral harvesting.

Underlying Causes

International trade

The growing international demand for traditional medicines, rare foods, curios and aquarium specimens etc. is leading to increasing over-exploitation of certain species within the region. The pressures on some individual species are illustrated by the value of their markets, for example, the market price for the body parts of a single tiger can be as much as US$5 million, and top-quality dried seahorses can sell for as much as US$1 200 per kilogram in parts of People's Republic of China. In addition, according to a recent report by TRAFFIC, a group that monitors the international wildlife trade, the annual Northeast Asian trade in tortoises and river turtles involves some 300 000 kilograms of live animals, with a value of at least US$1 million.

Traditional herbal therapies are also growing rapidly in popularity amongst industrialised countries of the world. The FAO estimates that between 4 000 and 6 000 species of medicinal plants are traded internationally, with People's Republic of China accounting for about 30 per cent of all such exports. In the early 1990s, the booming US retail market for herbal medicines reached nearly US$1.5 billion, and the current European market is thought to be even larger.

The Asian rattan palm is another species of plant that has been in decline in recent years due to high international demand to supply an international furniture making industry worth US$3.5 to 6.5 billion annually. Rattan stocks are being depleted at an unsustainable rate throughout tropical Asia due to a combination of the loss of native rainforest and over-harvesting. As a result, in the past few years some Asian furniture makers have even begun importing rattan supplies from Nigeria and other central African countries.

Population and poverty

Continued population growth and urbanisation is exerting a constant and degrading pressure on biodiversity throughout the region, primarily due to encroachment into natural habitats and their conversion for human settlement, or for the expansion of agricultural production to meet increased demand. In addition, the intensification of agriculture production is also encouraging the use of hybrid seeds and agricultural chemicals, thereby further degrading natural habitats and biodiversity.

Poverty is another significant underlying factor in regard to the loss of biodiversity in the region. The poor are frequently forced to occupy and/or subsist on marginal lands, and thereby often encroach upon fragile ecosystems, such as wetlands in Bangladesh, hill forests in India and Nepal and mangroves in Thailand. Urgent, but pragmatic, responses are therefore needed to address this problem, for example focusing on developing alternative livelihood strategies for those who currently rely on protected natural habitats for their living. It is worth mentioning, however, that it is by no means clear whether poverty, with its pressure to survive, or affluence, with its pressure to consume, ultimately leads to greater degradation of resources and the environment.

Bioinvasion

Bioinvasions are the processes by which new (or "exotic") species are introduced into a habitat. They are rated second only to habitat loss as the major cause of biodiversity loss in the region, where they are deeply enmeshed in basic economic processes such as trade and travel. Such processes have created hundreds of exotic "pathways" for invasion, for example, container shipments of used tires from Japan are believed to be responsible for the arrival of Asian tiger mosquito in New Zealand and Australia, and the ballast water of foreign ships was thought to contain the poisonous plankton which burst into the dinoflagellate "red tides" which devastated Australian fisheries in the 1980s.

Other examples include the introduction of the domestic cat to New Zealand, which is thought to have been the main factor in the loss of many bird populations in that country, and the invasion of "exotic" rats, particularly the black and the brown rat, which have also taken a similar toll on many island birds across the region. Another spectacular example involves the brown tree snake, which invaded Guam from Papua New Guinea around 1950 and has driven nine of the island's 18 native birds into extinction, along with several lizards and probably three species of bats. Although mammals have in general been less susceptible than birds to invasive species, a significant exception has been the unique marsupial and rodent fauna of Australia.

The introduction of non-native rabbits, foxes, cats, rats, and other animals in this country has combined with changing land-use patterns during the past two centuries to give Australia the world's highest rate of mammalian extinction. In total, 19 species of mammal have become extinct since European settlement in the Eighteenth Century, and at least one quarter of the remaining native mammalian fauna remains threatened. As yet there are few data to analyse their impact, but the rapidly increasing use of genetically engineered micro-organisms, and their establishment within natural habitats, could increase the potential risks to existing natural biota. This could occur, for example, through engineered genetic traits which are harmful to non-target species upon which other native biota depend, through a mixing of genetic stocks, or through their general competitive superiority and subsequent intrusiveness.

In agriculture, species which are considered favourable in some areas can become serious "biological polluters" in other some areas. For example, the spice cardamom is a problem in lowland moist forests of Sri

Lanka and southern India, whilst black pepper has invaded forest edges in Malaysia. In addition, Chromolaena odorata, the shrub valued by small farmers in Indonesia as a fallow crop, has become a serious pest for many major tropical crops, such as rubber, oil palm, and coconut, and arguably the single most invasive plant in tropical nature reserves.

Consequences of Biodiversity Loss

Impacts on the Ecosystems

One major consequence of biodiversity loss is the alteration and decline in species compositions, which may then trigger local and global extinctions. Evidence also suggests that removal of key herbivore and predator species may ultimately produce large-scale ecosystem changes. For instance, removal of triggerfish has been linked with explosions in burrowing urchin populations, which subsequently accelerate reef erosion through feeding activities. Moreover, the loss of a region's top predators or dominant herbivores is particularly damaging because it can trigger a cascade of disruptions in the ecological relationships among species that maintain an ecosystem's diversity and function.

Large mammals tend to exert inordinate influence within their ecological communities by consuming and dispersing seeds, creating unique micro-habitats, and regulating populations of prey species. Similarly, decades of excessive whaling reduced the number of whales that die natural deaths in the open oceans. This may have adversely affected unique deep-sea communities of worms and other invertebrates that decompose the remains of dead whales after they have sunk to the ocean floor. Loss of biodiversity through bioinvasion of exotics also has a serious repercussion as it could result in loss or alteration of genetic purity or genetic uniformity.

Exotics can pose a kind of internal threat to natives as they may cause the mixing of genetic stocks. An exotic closely related to a native may interbreed with it, releasing its genes into the native gene pool. Such genetic invasions can undermine the distinctiveness or stability of a native population by swamping it in foreign genes. Among plant species, many crops interbreed with wild relatives, and it is possible that these "exotic genes" could escape into wild plant populations—or that the crops themselves could escape. The appearance of herbicide tolerant or disease-resistant wild plants could obviously lead to major ecological- and agricultural-impacts. Mixing of genetic stocks could also occur in freshwater and marine species.

In marine ecosystems, biodiversity loss has two important genetic consequences. Firstly, it affects the genetic variability within a species. Species exhibiting broad genetic diversity are more likely to adapt to changing conditions than species with narrow genetic diversity. Population declines, by reducing genetic diversity, also reduce the ability of a species to adapt to changing conditions. Secondly, these losses can have cascading, unanticipated effects on other species within an ecosystem.

Impacts on Human

With the continued loss of terrestrial, freshwater and marine biodiversity in the region, fish, grains and other food and medicinal products which are derived from these ecosystems are also under increasing pressure. In most cases, these food and medicinal products are integral to the lives of poor and indigenous communities, and so it is they who are being forced to find alternative livelihoods, and who in many cases are suffering as a result. Moreover, although people who are more integrated into regional and national economies tend to use fewer natural resources, they still may depend on plant and animal diversity to generate cash income, for example in India, nearly six million people make a living by harvesting non-timber forest products, a trade that accounts for nearly half the revenues earned by Indian state forests.

Loss of biodiversity is also frequently associated with a decline in the quality of diet and/ or intake of food for the poor, which can exacerbate the incidence of malnutrition and sickness, especially amongst children. Moreover, humanity derives many of its medicines and industrial products from the region's wealth of biodiversity, and as plant and animal species with medicinal properties are lost, primary healthcare for millions of people across the region, and in particular the poor, is at risk. In response, professional ethno botanists surveying medicinal plants used by different cultures are racing against time to document traditional knowledge before it vanishes with its last elderly practitioners.

Manipulation of Genetic Viversity

The scale of manipulation of genetic diversity in recent years has been unprecedented. In particular, manipulation of pollens, seeds and other propagules through biotechnology and/or genetic engineering has contributed to changes in the patterns of occurrence of some genes. It has even resulted in transfer of genetic material from one species to another. This has brought revolutionary changes in agriculture, medicine and other fields. However, it has also brought new challenges for the protection and conservation of biodiversity in the region.

Environmental and Health Related Risks

The most important outcome of agricultural biotechnology to date is the production of genetically engineered plant seeds, more commonly known as genetically modified (GM) seeds. Genetic engineers claim that the new seeds are to all intents and purposes the same as the original ones, except that they have been improved to produce a higher crop yield, for example through increased resistance to pests, and enhanced response to particular fertilizers.

Critics of the new technology, however, assert that there are too many risks associated with genetic engineering, particularly in relation to environmental and human health. More specifically, they are concerned about biosafety of genetically engineered organisms, that is, their safe transfer, handling, use and disposal in the environment. In terms of environmental risks, the consequences of releasing genetically engineered seeds into the environment have not been adequately assessed.

Wind blown pollen from the modified seeds may, for example, fertilize related plants outside the growth area, and if it contains a gene for herbicide resistance, a new strain of "superweeds" could result. Similarly, it is difficult to predict with any certainty what effects the modified plants may have on other organisms. For example, genetic engineers have developed new strains of many plants that contain a gene from a bacterium Bacillus thuringiensis (Bt), which makes them resistant to insect pests. However, there is some evidence that the crops may have an impact not just on the pests, but also on beneficial insects such as bees, lacewing, and ladybirds and the birds that eat them. In terms of health risks, it has been found that certain transgenic crop varieties (e.g. Bt-maize) contain a marker gene that codes for antibiotic resistance in the bacteria Erwinia coli.

There is a risk that if animals or humans consume such products, for example in the form of cattle feed or starch, some antibiotics would be rendered useless. Another concern is that when crops such as rice or rapeseed with high Vitamin A concentrations are planted, there will be no way to distinguish them from normal crops, with the contingent risk of liver damage if too much Vitamin A is unwittingly consumed. Laboratory tests have also shown that certain genetically engineered materials are also associated with reproductive problems in rabbits.

In some countries, such as England and France, major supermarket chains have already banned genetically engineered products from their shelves, and in some Asian and Pacific Region countries, including Japan and Hong Kong, China, supermarkets have begun to introduce GM—free products to test consumers reaction. Reaction amongst some other countries is the biodiversity rich Asian and Pacific region, has also been hostile, for example, students from the Republic of Korea have blockaded government funded biotechnology greenhouses, and in India, farmers have made strong protests to the government. However, the reaction has been varied amongst different stakeholder groups throughout the region.

Intellectual Property Rights and Dependency: Intellectual property rights (IPR) is another area of major concern to some of the less industrialised countries of Asia and the Pacific which are rich in biodiversity; the worry being that they will be forced to accept or adopt some form of IPR, even before a full and frank international debate has taken place about the suitability of such systems to protect resources that are collectively owned. Moreover, if foreign researchers and transnational corporations can patent indigenous plants without adequately compensating the communities who provided them, there are fears that farmers will end up paying surcharges on products formed from their own knowledge and experience.

In response, some experts believe that IPR should also be applied in favour of the farming communities, to patent and protect their own varieties in order to retain any benefits, which are derived from them. However, this in turn gives rise to the problem of benefit sharing, how to define "community" and, in the case of plant breeding, whether it is possible to determine the provenance of a plant that is the product of generations of farmers' inputs. Probably the most important and challenging risk to biodiversity rich countries of the region in relation to genetic engineering is the balance of power over food supply, which is fast becoming concentrated in the hands of a few major companies who dominate the world's seed market, such as Monsanto, Novartis, Calgene, Du Pont and AstraZenica.

"Control of the plant gene" has aroused particular concern in this respect; a genetic technique which enables seed companies to ensure that farmers are not able to raise second generation (F2) populations from first generation (F1) seeds using mechanisms such as embryo abortion in the first generation seeds. Currently, 80 per cent of crops in the biodiversity rich world are grown each year from saved seed. The cost of buying expensive new seeds every year will clearly place a significant economic burden on many small farmers.

Bio-prospecting

Biodiversity prospectors search for genetic and biochemical resources that have a commercial value, particularly to the pharmaceutical, biotechnological, and agricultural industries. Biodiversity prospecting is not new—plant collectors from industrialised countries have ventured southward in search of valuable genetic material for agricultural plant breeding for many years—however, recent advances in molecular biology, and the increasing availability of sophisticated diagnostic tools for screening, have made it an increasingly cost-effective operation for pharmaceutical corporations and others to perform.

As a result, the market for buying and selling exotic biological specimens is expanding rapidly. The renaissance of bio-prospecting is also fuelled, in part, by the realisation that species, their genetic material, and the ecosystems of which they are a part are rapidly disappearing from the face of the earth. In the mid-1980s, pharmaceutical industry analysts warned that each medicinal plant lost in the tropical rainforests could lose drug firms possible sales of more than US$200 million.

Valuable chemical compounds derived from plants, animals and micro-organisms are generally more easily identified and of greatest commercial value when collected with the assistance of, or in combination with, the knowledge of indigenous peoples. For example, scientists have found that 86 per cent of the plants used by Samoan healers displayed significant biological activity when tested in the laboratory. However, the contribution of indigenous peoples has rarely been acknowledged or rewarded in the past, and no matter how convincing the rhetoric today, conservation and equity are likely to remain secondary issues for the various corporations involved in the practice of bio-prospecting. Moreover, once indigenous peoples share information or genetic material, mechanisms such as monopoly patents can mean that they effectively lose control over those resources forever, regardless of whether or not they are compensated.

8

Climate Change and Biodiversity

The world's climate patterns have always been highly dynamic. Recent decades, however, have seen new patterns, indicating more rapid change than has been seen for thousands of years. These changes have been linked to equally rapid changes in atmospheric chemistry, and the release of "greenhouse gases" by human activities.

For the first time in the history of our planet a single species is driving a significant shift in the global climate. International efforts are being focused towards reducing gas emissions. However climate change has already begun, and the impacts will increase considerably over coming decades, no matter how successful the efforts at emissions reductions.

Global biodiversity is inextricably linked to climate —changes in climate affect biodiversity and changes to natural ecosystems affect climate. Already hemmed in by habitat loss, pollution and over-exploitation, species and natural systems are now faced with the need to adapt to new regimes of temperature, precipitation and other climatic extremes. The rate and magnitude of climate change induced by increased greenhouse gases emissions has and will continue to affect biodiversity and might outweigh them in the future. Nature conservation in the new millennium has increasingly difficult challenges to face.

For a given ecosystem, functionally diverse communities are more likely to adapt to climate change and climate variability than impoverished ones. In addition, high genetic diversity within species appears to increase their long-term persistence. It must be stressed, however, that the effect of the nature and magnitude of genetic and species diversity on certain ecosystem processes is still poorly known. The ability of ecosystems to either resist or return to their former state following disturbance may also depend on given levels of functional diversity. This can have important implications for the design of activities aimed at mitigating and adapting to climate change. Therefore, conservation of genotypes, species and functional types, along with the reduction of habitat loss, fragmentation and degradation, may promote the long-term persistence of ecosystems and the provision of ecosystem goods and services.

Impacts of Climate Change on Biodiversity

Climate change is likely to have a number of impacts on biodiversity—from ecosystem to species level. The most obvious impact is the effect that flooding, sea level rise and temperature changes will have on ecosystem boundaries, allowing some ecosystems to expand into new areas, while others diminish in size. As well as shifting ecosystem boundaries, these changes will also cause changes in natural habitat—an outcome which will have a knock-on effect on species survival.

A growing body of research indicates that, as a result, climate change may lead to a sharp increase in extinction rates. Mid-range predictions from one recent study suggest that 24 per cent of species in the five study regions will be on their way to extinction by 2050 due to climate change. For many species, climate change poses a greater threat to their survival than the destruction of their natural habitat.

Global warming is also causing shifts in the reproductive cycles and growing seasons of certain species. For example, higher temperatures have led to an increase in the number of eggs laid by the spruce budworm, already one of the most devastating pests in North America's boreal forests. The impacts of climate change on biodiversity will vary from region to region.

The most rapid changes in climate are expected in the far north and south of the planet, and in mountainous regions. These are also the regions where species often have no alternative habitats to which they can migrate in order to survive. Other vulnerable ecosystems and species include small populations or those restricted to small areas. Coral reefs have already shown devastating losses as a result of increased water temperatures.

It is not just climate change itself that can have an impact on biodiversity. In some cases, the strategies that are adopted to mitigate climate change can affect biodiversity - both positively and negatively. Investment in renewable energy technology may provide climate change benefits, but outcomes for biodiversity are often poor. For example, some bio-energy plantations replace sites with high biodiversity, introduce alien species and use damaging agrochemicals.

Large hydropower schemes can cause loss of terrestrial and aquatic biodiversity, inhibit fish migration and lead to mercury contamination. They can also be net emitters of greenhouse gases if submerged soils and vegetation decay and release CO_2 and methane. By contrast, fuelwood conservation measures, such as efficient stoves and biogas use, can conserve carbon reservoirs and reduce pressure on forests. The concept of becoming 'carbon neutral' is gaining popularity with many businesses that wish to contribute to climate change mitigation activities by offsetting their carbon emissions. Likewise, many nations have committed to reducing their net greenhouse gas emissions under the Kyoto Protocol of the United Nations Framework Convention on Climate Change (UNFCCC).

Projects designed to sequester carbon, and hence mitigate climate change, present opportunities to incorporate biodiversity considerations. Afforestation and reforestation activities can restore watershed functions, establish biological corridors and provide considerable biodiversity benefits if a variety of different aged native tree species are planted. Monocultures, however, not only reduce biodiversity, but also increase the chances of pest attacks thus challenging the permanence of carbon stocks. The location of afforestation and reforestation projects is also important. Replacing native grasslands, wetlands, shrublands or heathlands may lead to dramatic biodiversity losses, and also lower the relative increase in carbon sequestered compared to implementing such projects on degraded land.

Impacts of Biodiversity on Climate Change

Just as climate change affects biodiversity, so changes in biodiversity can also affect the global climate. Land use changes that lead to biodiversity losses can cause increased greenhouse gas emissions. Forests are a major store of carbon, and when forests are cut down or burnt, CO_2 is released into the atmosphere.

Continuing deforestation, mainly in tropical regions, is currently thought to be responsible for annual emissions of 1.1 to 1.7 billion tonnes of carbon per year, or approximately one-fifth of human CO_2 emissions. Peatlands or mires hold roughly one-third of the carbon contained in soil worldwide, and greenhouse gases are released every time peatlands are burned, drained, converted to agriculture or degraded. Peatland forest fires in Indonesia in 1997 released an amount of CO_2 equivalent to 40 per cent of the world's average yearly carbon emissions from fossil fuels. Such peatlands also provide many environmental services, such as improving water quality.

Many are important biodiversity reservoirs or stopover points for migratory species. There are also feedback mechanisms at work between biodiversity and climate change. For example, some species of ocean algae release dimethyl sulfate (DMS) into the atmosphere. Rising ocean temperatures (a product of global warming) mean that more DMS is released from booming algal populations. But DMS is also associated with the formation of clouds, which may actually help reduce the amount of heat reaching the Earth's surface.

Effective biodiversity conservation and management can lead to higher levels of carbon sequestration and hence climate change mitigation. For example, forest management activities such as increasing rotation age, low intensity harvesting, reduced impact logging, leaving woody debris, harvesting which emulates natural disturbance regimes, avoiding fragmentation, provision of buffer zones and natural fire regimes, can simultaneously provide biodiversity and climate benefits. This is also true for certain agroforestry, revegetation, grassland management and agricultural practices such as recycling and use of organic materials. Integrated watershed management can conserve watershed biodiversity in addition to increasing water retention and availability in times of drought, decreasing the chance of flash floods and maintaining vegetation as a carbon sink.

Energy production is another key area where biodiversity conservation provides opportunities to help mitigate climate change. Currently, some 60 per cent of anthropogenic global greenhouse gas emissions originate from the generation and use of energy. Use of renewable energy sources provides an opportunity to reduce emissions from burning fossil fuels.

Impacts of Climate-induced Changes on Livelihoods

Poor people generally depend more on ecosystem services and products for their livelihoods than wealthy people. The means by which a poor family gains an income and meets its basic needs are often met by multiple livelihood activities. For example, exploiting common property resources such as fish, grazing land or forests can provide income, food, medicine, tools, fuel, fodder, construction materials, and so on. Poor people are therefore severely affected when the environment is degraded or their access to it restricted.

This link between poverty and the environment has been recognised for some time. As a result of this dependency, any impact that climate change has on natural systems threatens the livelihoods, food intake and health of poor people. Climate change will mean that many semi-arid parts of the developing world will become even hotter and drier, with even less predictable rainfall.

Climate-induced changes to crop yields, ecosystem boundaries and species' ranges will dramatically affect many poor people's livelihoods. Those most vulnerable to climate change are the poorest groups in the poorest countries of the world. This is because they live in areas more prone to flooding, cyclones, droughts, and so on and because they have little capacity to adapt to such shocks. They are often heavily dependent on climate sensitive sectors such as fisheries and agriculture, and the countries they live in have limited financial, institutional and human capacity to anticipate and respond to the direct and indirect impacts of climate change.

Conservation of biodiversity and maintenance of ecosystem integrity may be a key objective towards improving the adaptive capacity of such groups to cope with climate change. Functionally diverse systems may be better able to adapt to climate change and climate variability than functionally impoverished systems. A larger gene pool will facilitate the emergence of genotypes that are better adapted to changed climatic conditions. As biodiversity is lost, options for change are diminished and human society becomes more vulnerable.

Poor people are particularly vulnerable to extreme weather events. Over 96 per cent of disaster-related deaths in recent years have taken place in developing countries. Extreme weather events are increasing, and during 2001, 170 million people internationally were affected by disasters, 97 per cent of which were climate-related. Women and children are particularly vulnerable. For example, when the 1991 cyclone hit Bangladesh, 90 per cent of victims were women and children. This was due to a number of factors including their capabilities in survival, and socio-cultural beliefs that prevented women with their children from congregating in public cyclone shelters.

Classic top-down approaches to climate change equate to large infrastructure construction projects. Those designed to support adaptation to climate change are often associated with physical protection, for example large sea walls. Those designed to reduce global greenhouse gas emissions and thus mitigate climate change are often associated with large renewable energy schemes, such as hydropower. Such projects often have significant negative impacts on biodiversity and local livelihoods. For example, plans to build scores of dams with massive hydroelectricity generating potential on the Mekong River will affect the livelihoods of the 52 million people currently using river resources, many of whom live below the poverty line. Dam construction will prevent fish migration, and yet Mekong fish provide 40 - 60 per cent of the animal protein consumed by the population of the lower basin.

The nine proposed mainstream dam projects alone would also displace 60,000 rural people. While big infrastructure projects can be effective, comparatively little attention has been paid to nonstructural alternatives and to 'bottom-up' approaches rooted in existing community-based strategies for managing resources and reducing vulnerability to climatic shocks. Many of the carbon sequestration projects undertaken by companies and nations to mitigate their greenhouse gas emissions do not incorporate biodiversity issues. Even fewer incorporate livelihood and poverty issues as well.

The Clean Development Mechanism (CDM), established under the Kyoto Protocol of the Climate Change Convention, aims to provide developed countries that have accepted targets for reducing greenhouse gas emissions with flexibility for achieving these targets, by allowing them to take credits from emissions reduction projects undertaken in developing nations.

Projects are supposed to provide global benefits from carbon sequestration, but also sustainable development benefits to host developing countries. Many projects pay little attention to these sustainable development benefits. However, one project generating electricity from biogas and bio-diesel in Brazil is attempting to provide livelihood, carbon and biodiversity benefits.

Some initiatives succeed in combining biodiversity, livelihood and climate change related benefits. Activities to build the resilience of communities to climate stresses are continuing in Central America, where following Hurricane Mitch, the charity World Neighbours has been working to increase agroecological activities. This is helping vulnerable communities adapt to their changing environments as the incidence and severity of climate change-related disasters increases.

Similarly, since 1992, PASOLAC (Programa para la Agricultura Sostenible en las Laderas de América Central) has been helping communities in Nicaragua, Honduras and El Salvador to increase the agricultural productivity of their hillsides through improved soil and water management. The programme is characterised by participatory demand driven approaches. Such hillsides support tropical forests, key freshwater reservoirs, and several important and diverse ecosystems. They also represent the economic base for the majority of the population in Central America (producing grains, coffee etc.) but are prone to soil and landscape degradation, problems exacerbated by climate change induced droughts and floods. 'Bottom-up' processes have also met with success in Bangladesh, through the Reducing Vulnerability to Climate Change (RVCC) Project and Vietnam, where the Vietnam National Chapter of the Red Cross has worked with local communities to rehabilitate mangroves.

Climate Change, Biodiversity Conservation and Poverty Reduction

Climate change is dealt with in the Millennium Development Goals (MDG) on ensuring environmental sustainability. One target set for reaching this goal to 'integrate the principles of sustainable development into country policies and programmes and reverse the loss of environmental resources'. Indicators for monitoring whether this target is met are as follows:

— Proportion of land area covered by forest

— Ratio of area protected to maintain biological diversity to surface area

— Energy use (kg oil equivalent) per $1 GDP (Purchasing Power Parity)

— Carbon dioxide emissions per capita and consumption of ozone-depleting Chlorofluorocarbons (Ozone Depleting Potential tons)

— Proportion of population using solid fuels

The focus on energy use and CO_2 emissions emphasises mitigating climate change and ignores the fact climate change is already a reality, and adaptation needs to be considered. This need for adaptation does not replace the need to mitigate climate change; both adaptation and mitigation are important parts of the solution. The proportion of population using solid fuels is also a poor measure of air quality as it fails to distinguish between the dangers of indoor and external air pollution.

The forest indicator would be better if it captured some measure of goods and services coming from forests, such as carbon sequestration, soil protection, biodiversity enhancement and contributions to local livelihoods. Likewise, the indicators relating to energy use and CO_2 emissions would be improved if they captured some measure of broader environmental and social benefits emerging from activities undertaken. If such considerations are taken into account, investment could shift towards projects with multiple livelihood, biodiversity and climate change benefits (or at least projects which do no harm in these additional contexts), as opposed to initiatives (such as large dams), which might meet one indicator, but which have significant negative impacts on ecosystem integrity, biodiversity, climate change mitigation or adaptation, and local livelihoods. Achieving the MDGs is already proving to be a challenge. Ensuring that any progress made towards meeting these targets will benefit the poor depends on the process undertaken.

The role of local processes in addressing climate change, biodiversity conservation and poverty reduction is often overlooked - or undermined. The biodiversity conservation agenda has, for example, been driven by international priorities - such as preservation of rare and charismatic mammals. This approach has often superseded local values attached to biodiversity such as livelihood support or risk reduction in the face of climate shocks.

Likewise, much attention and funding, currently focuses on top-down strategic planning requirements such as National Communications, National Adaptation Plans of Action, National Biodiversity Strategies and Action Plans, National Conservation Strategies, National Environmental Action Plans and Poverty Reduction Strategy Papers. Such initiatives typically challenge poor countries by placing considerable strain on already overloaded institutions with limited capacity.

Supporting local solutions requires action at several levels. Globally, actions to reduce poverty and inequity will ultimately reduce vulnerability to climate change and may also reduce unsustainable natural resource use. Such actions include curbing the loss of income from trade barriers and subsidies paid to farmers in high income nations; improved market access for processed raw materials; external investment and untied aid (to support accountable local processes); debt relief; and a commitment in high-income nations to change consumption patterns and thus reduce greenhouse gas emissions.

Synergies between the Climate Change Convention and Convention on Biological Diversity need to be explored, alongside links with development plans. This is not easy as the processes have separate constituencies, administration arrangements, negotiators and guiding scientific bodies. Some feel encouraging countries to establish a single body to deal with their obligations under all international environmental agreements would be useful. For example, disaster management plans drawn up to deal with climate induced disasters could incorporate impacts on local ecosystems in addition to vulnerable human settlements. This would recognise the role that ecosystems play in local livelihoods as well as havens of biodiversity.

Possible tools for integrating biodiversity, livelihood and climate change concerns include the ecosystem approach, which could incorporate climate concerns and environmental assessments which can be adapted to support broad uptake of environmental, social and development priorities. Measuring the value of environmental services to capture the true value of environmental goods and services is needed. Participatory processes and an holistic approach incorporating all aspects of sustainable development should be promoted.

Nationally, policies that benefit biodiversity, climate change adaptation and mitigation, and poverty reduction need promotion. In particular, development activities should integrate responses to climate risks and thereby minimise the impacts of climate change. Development agencies, national governments and other stakeholders should internalise climate change into their work. However, many adaptation activities are located within the Ministries of Environment, which are traditionally relatively weak and have little influence over line Ministries. Good governance is very important, but inevitably requires contextualisation within the complexities of local and national political systems. Stronger decentralised government can play an important role, but a well functioning national government with vision and accountability is also critical.

9

Impacts of Dams on Freshwater Biodiversity

Biodiversity is considered at four levels: genetic; species; ecosystem and ecological function. Normally it is the indigenous or native components of biodiversity that are examined; exotic or alien species are a separate component, although interacting with the native species. Many species are yet to be discovered, scientifically named and classified, especially in tropical regions and in some taxonomic groups that are poorly studied, for example the nematodes, algae, bacteria and fungi. Perturbations to a species status can be measured in simple terms by reduced population sizes, extirpations, or extinction.

Conserving habitats and ecosystems is the key to species conservation. Here habitat may be defined as the place where an organism lives or living space and an ecosystem as the interaction or functioning between a community of organisms and their nonliving environment. The earth can be divided into a series of biogeographical regions, or biomes, ecological communities where certain species of organisms co-exist within particular climatic conditions. Within a biome there are several local factors which affect the distribution of species.

Degradation of habitat leads to lowered population size and loss of habitat to extirpation. Humans mainly induce extinctions by causing habitat loss. The importance of all components of the ecosystem including primary producers, herbivores, carnivores, detritivores and recyclers and their ecological function should be considered in the design of dams. In addition there are certain special linkages between species, e.g. freshwater mussel larvae (glochidia) are parasitic on the gills of fishes for part of their life cycle. In many cases specific fish hosts are required.

Assessing the status of biodiversity at the global level is difficult because the evaluation is incomplete and uneven. Birds are quite well assessed, oligochaetes poorly, and many microbiota not at all. The IUCN Red Lists ease the task by bringing together what is known and applying uniform criteria. However, summaries do not separate freshwater biota from biota in other environments. Table 1 summarises terrestrial and freshwater vertebrate data at the global level.

Table 1. Proportion of terrestrial and freshwater vertebrates globally threatened.

Group	*Proportion Threatened (%)*
Mammals - all	25
Land birds	11
Waterfowl (freshwater)	13
Turtles, tortoises and terrapins	38
Crocodiles	43
Amphibians	25 (estimated)
Freshwater fishes	33 (estimated)
Freshwater mammals (freshwater dolphins and otters)	65

The level of threat for dominantly terrestrial vertebrates is 11 to 25%, while the remaining values for groups occurring more frequently or uniquely in freshwater range from 13 to 65%. This gives a sense that, globally, freshwater species are more at risk than terrestrial species. Waterfowl are more threatened, 12.7%, than land birds, 10.8%. Similarly, freshwater mammals are more threatened, 65%, than all mammals, 25%.

Molluscs: Extinctions are a significant problem in terrestrial, freshwater and marine molluscs. Of all the species that became extinct since 1600 AD, 37% were mollusc species, which is more than any other group evaluated. These percentages refer to the total known globally. It should be made clear that the percentages are affected by the degree to which the group has been studied and the number of species in the group, e.g. the status of birds is better studied than that of molluscs, but there are more mollusc than bird species.

The 1996 IUCN Red List of Threatened Animals lists 12 bivalves and 216 gastropods as extinct, and 114 bivalves and 806 gastropods as threatened, for a total of 228 extinct and 920 threatened terrestrial, freshwater and marine molluscs. About 18% or 145 of the threatened molluscs are spring molluscs. Data on threatened freshwater molluscs are given in Table 2.

Table 2. Freshwater molluscs in the IUCN 1996 Red List of threatened animals.

Risk category	*Bivalves*	*Gastropods*	*Total molluscs*
Extinct	12	14	26
Critically endangered	85	60	145
Endangered	24	86	110
Vulnerable	8	194	202
Near threatened	66	35	101

Data deficient	4	104	108
Total listed	199	493	692

Spring snails in the critically endangered category in Austria, Australia and United States are threatened by over-abstraction of water from their habitat or by pollution. African freshwater molluscs are threatened by decline in quality of water, pollution, damming, and increased sediment load.

Wilcove et al. reviewed the threats to imperilled species in the United States. They found that the chief threats to freshwater molluscs were habitat degradation and loss, 97%, pollution, 90%, and alien species, 17%. The chief causes of habitat loss were water development, 99%, pollution, 97%, dams and other barriers to flow, 96%, and agriculture 64%. More recently in North America, alien zebra mussel invasions have become the major factor in loss of native mussel diversity.

Fishes: The 1996 IUCN Red List of Threatened Animals lists 617 freshwater fishes; about 6% of the known number of freshwater species. The Red List has evaluated only a fraction of freshwater fishes, therefore a conservative estimate gives 20% as extinct, endangered or vulnerable, or more realistically 30-35%. In the United States, freshwater fishes have been threatened by water development (91%), dams, impoundments and other water barriers (64%), pollutants (55%) and agriculture (45%).

Birds: According to the 1996 Red List, 1,107 species or 11% of all bird species are threatened and 104 are extinct. Amongst the more threatened of bird groups are the aquatic Gruiformes (rails and cranes) with 54 species, and the partially aquatic Coraciiformes (kingfishers and bee-eaters) with 11.5% threatened, while 18% of the Podicipediformes (grebes) are threatened. Extinct aquatic birds include the Colombian grebe (Podiceps andinus) and the Atitlan grebe (Podilymbus gigas). Thirteen per cent of globally threatened birds are (freshwater) water-birds. There are 90 species of critically endangered, endangered and vulnerable water-birds. This suggests that water-birds are slightly more threatened than land birds. Habitat loss and degradation are the key factors affecting threatened species. This is generally due to the loss and change through drainage and land reclaiming, converting natural wetlands into urban and industrial lands in Europe and into agricultural land in North America.

Plants: There are about 270,000 scientifically described species of vascular plants, but the true number may be in the order of 300,000-350,000 species. Two important documents on global plant biodiversity have been published in recent years, Centres of plant diversity, in three volumes, and the 1997 IUCN Red list of threatened plants. About 78% of the world's plants are tropical. More than 40,000 plant species, about a quarter of the world's tropical plant diversity, occurs in Colombia, Ecuador and Peru, while Brazil has between 40,000 and 80,000 species. Tropical and subtropical Asia has at least 50,000 species and Southern Africa has 21,000 species of plants, of which 80% are endemic.

The Red list of threatened plants demonstrated that 32,112 species or 11.9% of the world's 270,000 vascular plant species, are threatened, and 374 or 14% are extinct. The counts are for terrestrial and aquatic species combined. Of these, 6,522 species are classed as endangered. Thirty-two countries have at least 5% of their native species threatened. The main countries having high proportions of threatened species, 11-29%, include the USA, Jamaica, Turkey, Spain, Australia, Sri Lanka, Cuba, Panama, Japan and Greece. In the United States about half of the potentially extirpated species are either obligate or facultative native wetland species. These wetland plant species may be affected by changes in the aquatic environment mediated by dams.

Terrestrial species may be affected by reservoir flooding, de-watering downstream of dams; construction of transmission lines, access roads or canals or through lowered water tables. Walter and Gillett list the numbers of species, the number globally threatened by category and the threatened percentage for each family of vascular plants. The status of selected freshwater plants based on their Red List is presented in Table 3.

Table 3. Status of selected families of freshwater plants

Family	*% of threatened species*
Water lilies, Ceratophyllaceae	16.7
Water lilies, Nymphaeaceae	8.0
Water poppies, Limnocharitaceae	31.6
Water plantains, Alismataceae	13.3
Reeds and sedges, Cyperaceae	7.0
Frog-bits, Hydrocharitaceae	14.0
Duckweeds, Lemnaceae	9.7

This small sample illustrates the level of threats, 7.0-31.6% in 'water-loving' plants is about as high as in all vascular plants (terrestrial and aquatic) 11.7%. The mean, 14.3, and the midpoint, 19.1, of this sample are both higher than the mean of the family values for terrestrial and aquatic plants combined. This suggests that freshwater plants are more threatened than land plants. About half of the world's wetlands have been lost over the past century. In Asia more than 5,000 km^2 of wetland is lost every year due to agriculture, irrigation, dam construction, etc.

Status of Freshwaters

The impacts on the freshwater environment including:

— River seasonal flow patterns (levelling)

— River flow volume (reduction)

— Accessibility of species to river segments (blockage of migrations)

— Input of organic matter (leaves, wood, insects) (reduced)

— Toxicity (increased)

— Turbidity & sediments

— Nutrient levels (increased).

Freshwaters, and especially rivers and wetlands, are amongst the world's most severely impacted and have received many of the direct effects of human activities. Long-term and quality data on river discharge patterns are poor. Historical hydrological data are rare for many rivers. However data from old air photos and satellite images, including the new Radarsat images, which can be made through clouds, offer a promising source of information in particular the extent of flooding in the river floodplains. An analysis by Postel indicates that humans presently use 54% of the geographically and seasonally accessible runoff.

Demands by the year 2025 may increase to more than 70% of accessible runoff. The existing and growing friction between countries over shared waterways shows that humans are already facing water shortages that are international in scope. Problems already exist at the national and local level between and within sectors such as agriculture, domestic users and industry. The fixed supply of water poses profound problems for how it will be shared by aquatic life and increasing demands of humans.

Large Dams

Large dams are usually >15 m from foundation to crest. Dams of 10-15 m can also be defined as large dams if they meet the following criteria: crest length 500 m or more; reservoir capacity of at least one million cubic metres; maximum flood discharge of at least 2,000 m^3s^{-1}; 'specially difficult' foundation problems, or 'unusual design.' Major dams meet one or more of the following criteria: at least 150 m high; having a volume of at least 15 million m^3; reservoir capacity of at least 25 km^3; or generation capacity of at least one gigawatt. There are 306 major dams in the world and 57 are planned in the near future.

Construction of large dams includes the creation of access roads, preparation of the reservoir, excavation, construction of buildings and dams within and between river diversions, digging of canals and erection of power lines. Forested reservoir basins provide a particular challenge. Leaving some trees may provide fish habitat although leaving trees in any quantity may pose problems for future fishing, water quality and turbine safety.

While most reservoirs tend to trap sediments, e.g. a new delta is being formed within Lake Nasser, in some cases e.g. South Indian Lake Manitoba, the exposure of clay soils to shifting reservoir water levels increases erosion and downstream sediment discharge. The quantity of water discharged, when it is discharged during diel and seasonal cycles relative to the river's natural flow pattern and abiotic characteristics of the discharge such as temperature, oxygen, turbidity, and water quality significantly affect downstream biodiversity.

Globally terrestrial and aquatic ecological functions have been calculated to be minimally worth US$33 trillion per year, almost twice the value of the global gross national product, some $18 trillion, although the figure contains the value of some biological resources as well as functions. Costanza et al. indicated that the annual per hectare total global flow value of inland water systems, US$6,579 x 10^9 exceeded that from all other non-marine ecosystems combined—US$5,740 x 10^9.

Ecological functions, although not ordinarily included in gross global or national/domestic products nevertheless make significant contributions to economies. Freshwater ecosystems are economically more valuable than terrestrial ones. In many developing countries, fishes, including those from freshwater make a notable contribution in animal proteins to an otherwise carbohydrate-based diet. In the Amazon, the per capita consumption rate is 67 $kgyr^{-1}$ higher than in many areas.

In Tonie Sap, Cambodia, 100,000 tonnes of freshwater fish are caught annually, which source alone would provide a per capita 10 $kgyr^{-1}$. Biodiversity has many kinds of values and potential benefits for humans and the world as a whole. Before it is diminished, those responsible may well wish to consider the Precautionary Principle and take action to conserve it before components of it are permanently lost, even when the evidence for loss is not as strong as might be desired. That approach is advocated by the Convention on Biological Diversity.

Patterns of Freshwater Biodiversity

Although freshwaters comprise only 0.8% of the surface area of the world and they have fewer species than other systems, Table 4 shows that freshwaters contain more species per unit area than terrestrial and marine environments. This is particularly notable for fishes.

Table 4. Species richness of the world's major environments

Environment	*Area of world*	*No. living species %*	*Richness: % species %area*
Freshwater	0.8	2.4	3.0

Terrestrial	28.4	77.5	2.7
Marine	70.8	14.7	0.2
Symbiotic	N.A.	5.3	N.A.

Global biodiversity patterns in the three study groups, molluscs, fishes and vascular plants is presented in Table 5. Transitional ecosystems such as estuaries and land-water interfaces are not presented.

Table 5. Number of species per environment in study groups (world counts).

Group	*Total species*	*Land species*	*Freshwater species*	*Marine species*
Molluscs	95,500	24,500	5,000	65,500
		26.0%	5.0%	69.0%
Fishes	24,618	0	9,966	14,153
		0.0%	40.5%	57.5%
Birds	10,100	9,043	840	217
		89.6%	8.3%	2.1%
Vascular Plants	270,000	194,400	75,600	
		72.0%	28.0%	<1.0%

The total number of species (species richness) is only one measure of biodiversity. Other measures include species abundance. For example molluscs can form significant proportions of the benthos; 80% of the biomass of the River Thames at Reading is composed of freshwater unionid mussels. In the mountains, the mean fish biomass varies from 27-80 kg ha^{-1}, in foothill streams from 90-500 kg ha^{-1} and in the lowland streams from 300-600 kg ha^{-1}, while in human-made lakes it varies from 65-200 kg ha^{-1}. Diversity indices, which measure the relative richness, evenness, rarity and abundance, require quantitative sampling and at present there are little data for aquatic ecosystems.

Longitudinal gradients

Species richness can change from the headwaters to the river mouth. This may be related to changes in stream order, water temperature, oxygen, current, turbidity and available nutrients. The small headwater streams may have low numbers of fishes that increase downstream as the number of available habitats increases. Further and larger downstream segments of the river may have moderate numbers of species because lower habitat variety, and river estuaries, where salinity varies, may also have moderate species numbers.

Higher numbers of fish species are found in the Tennessee-Cumberland plateau drainage of the Mississippi, USA, than in the adjacent mainstream Mississippi; this may reflect more numerous habitats in the varied topography of the former, compared to the more constant gradient of the latter. Different species may characterise different river segments. Trout (*Salmo*) and sculpins (*Cottus*) may occupy headwaters, minnows and catfishes (*Cyprinidae and Ictaluridae*) midriver sections, and euryhaline species are found in the estuaries. Freshwater molluscs generally increase from the headwaters to the river mouth. This again is related to an increase in habitats in the floodplain areas in the middle or lower reaches.

Latitudinal gradients

Latitude-longitude grids have the disadvantage that the size of their grid cells shrinks towards the poles, a disadvantage in making comparisons of numbers of species if the study area spans several degrees of latitude. It is generally accepted that the number of species tends to increase from the poles to the tropics. Arctic waters can therefore be expected to contain fewer species than ones in the tropics.

In many freshwater lakes in the Arctic there is only one fish species, the Arctic charr, Salvelinus alpinus. The greatest freshwater fish diversity is found in Southeast Asia, tropical South America and central Africa although many fishes have not been scientifically described. Brown demonstrated the latitudinal trends in African freshwater molluscs which decreased in diversity from the tropics towards the Mediterranean and southern Africa.

Arid areas

Species diversity of North American fishes was related to a measure of 'aridity'. Data indicated that more species were found in moist compared to arid areas. The study showed at what critical level of moisture fish diversity began to increase rapidly. Although arid areas may be poor in species, desert springs or other water bodies may be rich in endemic organisms, for example, there are vast numbers of endemic spring-snails and fairy shrimps in the arid west regions of the USA.

Hotspots areas

'Hotspots' are geographic areas rich in species. 'Hotspots' often dominate over latitudinal and other gradients as shown for plants and animals. McAllister et al. listed countries that are rich in absolute numbers of aquatic species and in the number of species per unit area. There is some evidence that 'hotspots' for different groups of freshwater organisms are correlated, for example McAllister et al. found that the numbers of amphibian and fish species in 12 'mega-diversity' countries were highly correlated. Revenga et al. catalogued the number of fish species, endemic bird areas and percentage area of wetlands for most of the world's primary watersheds.

Global 'hotspots' of freshwater mollusc species include the Mobile Bay and Tennessee River basin faunas in the United States; the lower Mekong River of southeast Asia; the northern Western Ghats, India; the Lower Uruguay River and Rio de la Plata; lower Zaire; Lake Tanganyika; Balkans region; Lake Baikal. Note that the freshwater molluscs display very high levels of endemism. Global, regional and national 'hotspots' are often the dominating feature in geographic patterns of biodiversity. Hence it is vital that they should be taken into account in evaluating prospective dam sites. However more accurate identification of freshwater 'hotspots' is needed. If a standardised method was used for different groups it would assist in identifying common underlying factors and assist in application of the data to environmental impact analyses.

Phylogenetic and Ecological Diversity

Species are the most common units used in evaluating biodiversity. Yet species are only one unit in a hierarchy: Since families are more distinct genetically, ecologically and behaviourally, and generally have a more ancient origin than species, many taxonomists and conservationists would give a higher priority to conservation of the sole species representing a family than to one of a family with numerous species.

Alien (exotic) Species

Disturbed environments created by dams can foster populations of alien species and that diversions associated with dam projects may enable the invasion of such species. In the environmental assessments of dams, alien species should be separated from indigenous ones, and not tabulated in measurements of local indigenous biodiversity. In general, dam projects should not foster the introduction of alien species.

Influence of Dams on Biodiversity

There are a number of different migratory patterns of river-dwelling species. These include the well-known anadromous fishes e.g. salmon and hilsa and the catadromous fishes such as eels. Adults of anadromous species migrate up rivers to spawn and the young descend, while the reverse occurs with catadromous species. But many other freshwater fishes move up rivers or their tributaries to spawn, while the glochidia larvae of freshwater mussels 'hitch rides' on host fishes. To help counteract the drift downstream of their larvae, some aquatic insect adults such as mayflies and stoneflies fly upstream to lay their eggs. Dams block these migrations to varying degrees. However most waterfowl are able to fly over dams.

Reservoirs provide waterfowl habitat and may aid longer migrations by providing 'stopover' sites. The blockage of fish movements upstream can have a very significant and negative impact on fish biodiversity. Many stocks of Salmonidae and Clupeidae have been lost as a consequence. In the Columbia River, USA, more than 200 stocks of anadromous, Pacific salmonids became extinct. Sturgeon populations in the Caspian Sea rely on hatcheries, mainly in Iran, since Russian dams block natural spawning migrations. Hydroelectric dams in the Amazon basin have halted the long distance upstream migration of several species of catfishes and interrupted the downstream migration of their larvae.

On the Araguaia-Tocantins River basin, Brazil, several species of migrating catfish have been drastically reduced in abundance as a result of dams; catches in the downstream fisheries have been reduced by 70%. Diadromous fishes those that migrate at regular phases of their life history between freshwater and the sea, comprise about 250 species, <1.2% of all fishes species but form 3% of those classed as endangered. He observes that amongst them are species of great importance to fisheries, out of proportion to their number. Due to their occupation of connected habitats through which passes are needed at two or more life history phases, they pose special problems for conservation.

In particular the diversity of habitats used, the extensive areas occupied, the spatial separation of the habitats and the need for fish passage between them. Fishways must be designed to assist not only upstream and downstream migrations of large, fast-swimming migrants such as salmon that can pass substantial barriers, but also those of the lesser-known climbers like eels and gobies that require continuous dampened surfaces on which to move.

In general, a river is a one-way system for molluscs, as many molluscs can only move downstream by drifting or being dislodged by flood events and moved downstream. But some species with a larval form can move significant distances upstream with the aid of a third party, e.g. host fish during the larval stage.

Transfer of Matter Downstream

Nutrients: Conventionally, rivers have been regarded as the one-way transfer of matter downstream. Experiments by the Fisheries Research Board of Canada several decades ago showed that removal of spawned out sockeye salmon carcasses from streams reduced the growth of fry in the following year. Recently there has been greater appreciation that migrating species carry nutrients upstream. Reimchen proposed that intensive coastal catches of Atlantic salmon in the Queen Charlottes, Canada reduce the nutrients in adjacent stream riparian zones and estuaries.

The contribution of nutrients from both Atlantic and Pacific salmon carcasses has been linked to riparian tree growth. Data from 45 watersheds in British Columbia suggests that up to half the nitrogen stored in giant oldgrowth trees originates from sockeye salmon, using the nitrogen isotope N_{15} as a tag. The largest source of N_{15} is in the ocean. Cederholm et al. reviews the contribution of Pacific salmon carcasses to the flow of nutrients and energy for aquatic and terrestrial ecosystems.

Anadromous fishes carry nutrients in their bodies as well as gametes up river. The fishes and their eggs are used for food by a variety of aquatic and terrestrial predators, scavengers and detritivores. Decaying bodies, eggs and faeces of the consumers provide nutrients for the algae and other plants. The decreased production could be selfperpetuating, since small stocks produce lower amounts of in-stream nutrients for themselves, as well as other species. To a degree fertilisation from the lower riparian vegetation will also affect fish productivity.

The fall of insects, leaves and twigs into streams, which serves as direct or indirect food can be expected to decline, as riparian vegetation growth diminishes. There is no reason to suppose that the upstream transport of nutrients is restricted to anadromous salmonids. It can be expected that andromous hilsa in southeast Asia, Arctic whitefish of the genera *Coregonus and Stenodus*, lampreys such as Petromyzon marinus, and New Zealand retropinnids such as *Stokellia anisodon* would also transport nutrients upstream during their spawning migrations.

Turbidity: Reservoirs trap suspended particles, reducing turbidity downstream. Many species are adapted to natural turbidity. For example turbid water catfishes have small eyes, refined senses of smell and touch in their sensitive barbels. The turbid water helps conceal the fish and other biota from visual predators like birds. When normally turbid water becomes clear below dams, the indigenous species may find themselves at a disadvantage. Other animal species may move in, filter feeders and aquatic vegetation may flourish.

Sediment burrowing species may find their habitat has diminished. Flood plain ecosystems and deltas may no longer be replenished by the annual transport of sediment. Silt and increased turbity, above natural levels, can interfere with primary production. In the Mekong River system, silt levels increased following deforestation. This resulted in siltation of the river, lakes and swamps threatening the river fisheries.

Large organic debris (LOD): This consists of branches and tree trunks that fall into the river because of age, storms, beaver activity and eroded banks. Numerous organisms feed on LOD which is often the first link in the food chain. Trees can also play a complex role in creating habitats e.g. they divert, slow and speed up current flow, they shelter a variety of biota from currents and predators and create feeding stations. On land, LOD helps stabilise slopes and reduces erosion, and is converted into humus which helps hold water and moderates the runoff.

In estuaries, along shores of lakes and coastal areas LOD functions as a source of food, energy and habitat. In the Santilla River, Georgia wood represented 4% of the total habitat, yet supplied 60% of the invertebrate biomass and 78% of the drifting invertebrates. Dams tend to 'sieve out' LOD. The logs and branches may become waterlogged and sink, drift onto the shore or are removed by booms or other systems designed to protect turbines. If the integrity of downstream ecosystems is to be maintained, then LOD input must be sustained.

Lateral Impacts

Species and materials may move laterally away from the river, extending the effect of river changes to a band of varying width, parallel to the river.

Watering wildlife: As long as there is sufficient river flow below the dam, wildlife such as deer, antelope and elephants will come to the water, especially in the dry and hot season for drinking. Hippopotamuses will use water of sufficient depth as a day-time refuge, emerging to forage at night. Many birds may fly in to drink. These lateral movements can extend to several kilometres from the river. The reservoir itself, however, may serve as a source of water during the dry season or droughts, to wildlife living within range.

Watering terrestrial vegetation: Water release protocols can lower water tables lateral to the rivers which may affect vegetation there. According to LaRoe riparian ecosystems along most major western rivers of the U. S. have changed as a result of water development and flood control. Losses of riparian forest downstream of dams have been reported throughout western North America. Cottonwood-willow stands are being replaced by nonnative woody species such as Russian olive and tamarisk. This may result in diminished LOD input.

Loss of riverbank forests: The extinction of species may be related to the loss of gallery forests adjacent to rivers which became submerged following dam construction. The landsnail, Anthinus albolabiatus, was formerly endemic to gallery forest adjacent to the Uruguay River but became extinct after the formation of the Salto Grande Dam, Uruguay. This factor is of concern with regard to proposed dam projects in South Africa, where much of the remaining forest is preserved on the steep sides of valleys, which are also suitable sites for dam construction.

Maintaining populations and gene flow: The main stem of a river performs two related ecological functions to biota in tributaries. Periodically, populations of a tributary stream species, particularly fish, may go extinct and may be restocked from nearby rivers. Secondly individuals, as described above, may ascend a non-home tributary and contribute to the resident population's genetic diversity. For example salmonids are known to home to their natal streams for spawning, using celestial and olfactory cues. A small percentage of migrants are known, through tagging studies, to stray into adjacent tributaries or rivers.

Straying can serve to repopulate streams and also contribute to the genetic diversity of populations through gene flow. A single dam and more significantly multiple dams along a given river interfere with the genetic bridging function of the mainstem. Thus dams on the main stem can influence species diversity in lateral tributaries, even though there may be no changes in water flow or quality characteristics. In the Murray River, Australia, river regulation has lead to the separation of billabongs from the main channel habitat and thus for many molluscs in particular freshwater gastropods. Since river regulation was introduced some mollusc species have become extinct.

Oxbows, wetlands and springs: Oxbows, ponds, lakes and wetlands are often isolated in the floodpain from the river's main stem. These may be replenished with water, biota, sediment and nutrients during natural, seasonal floods. Levelling out of river discharge has been known to prevent these periodic linkages with the mainstem. Most of the 50 species of fishes in the Austrian Danube, depend on the connection between the river and its backwaters.

Transmission lines: Transmission lines affect biodiversity on land. The use of herbicides to control plant growth under power lines probably reduces native plant diversity in favour of weed species, which are often exotic.

Baker reported that power lines corridors can serve as refuges for rare species. Tree trimming may provide a haven for native 'sun-loving plants'. Shrub communities may flourish under power lines and provide habitat for nesting and migratory birds. Only nine snoutbean plants were thought to exist in all of Kentucky, USA.

However an East Kentucky Power Plant Corporation (EKPC) study showed that about two thousand specimens survived under their power line. EKPC adjusted their mowing schedule and removal of woody plants and educated other utilities about protecting native plants. Utility rights of way may harbour rare birds, amphibians, reptiles, tree snails, mammals and other species.

Upstream Area

Above the reservoir: Water quality, flow and seasonality of flow are not normally disrupted in the upstream area above the reservoir so impacts are generally less than for the reservoir and downstream areas. Nevertheless, the dam and the reservoir affect migratory movements of species into and out of this upstream area. The genetic exchange with downstream segments is reduced or prevented.

Reservoirs: In the construction of reservoirs, the clearing of vegetation, movement of earth and rock, the presence of humans and machinery, bringing in construction materials, use of explosives, noise, and reducing or cutting off river flow and increasing turbidity, will affect biodiversity. Removal of forests or other vegetation over a wide area, excavation, earth and rock movement and reduction in river flow are the most significant. Some of the on-site activities are mirrored in off-site disturbances such as the mass displacement of earth and rocks and road building. During reservoir filling the river and any associated wetland areas become inundated.

Riffles, runs and pools of the river are lost beneath the rising waters, leading to the extirpation (or extinction) of habitat sensitive riverine species with tightly defined niche requirements. Fishes in rivers are generally well adapted to flowing water. Similarly molluscs are often restricted to specific habitats within the river system, e.g. some species are bottom-dwelling filter feeders, others live in weeds at the edge of the channel. The transformation of a river to a reservoir therefore poses a problem for the resident, mainly riverine species that are not adapted to the new conditions.

Lacustrine fishes have been introduced into reservoirs in a number of cases e.g. Lake Kariba although this may pose new problems. In eastern Canada, lakes and streams, which have emerged fairly recently from glaciation, contain a number of fish species able to dwell in both habitats. However some species like the longnose dace, Rhinichthys cataractae, and rainbow darter, Etheostoma caeruleum, are adapted to running water only.

Molluscs generally show a drop in species richness from pre- to post-impoundment. Unionid mussels are exposed to a number of changes in the reservoir impoundment. The most important ones include: changes in the fish host population size (needed for the glochidia stage of the mussel larvae), reduction in the mussel population size (fertilisation success), eutrophication (affecting adult and larvae) and dissolved oxygen levels (low levels of oxygen in the profundal zone would eliminate mussels from that zone).

Eutrophication can limit the interstitial habitat of post-glochidial juveniles. Raised nitrate and phosphate levels are especially deleterious to juveniles. Organic debris can clog benthic interstitial spaces. Some of these effects can extend downstream of the dam. Table 6 shows the environmental and biotic factors which influence the various stages in the life cycle of unionid mussels.

Table 6. Factors affecting the life cycle stages of unionid molluscs. Items highlighted in bold are directly impacted and those in italics indirectly.

Factor	*Life cycle stages affected*
Exploitation	Adults
Fish host stock size	Glochidium
Mussel stock size	Fertilisation, glochidium numbers
River bed	Adults, juveniles
Flow regime	Fertilisation, glochidium infection, settlement, juvenile and adult
Suspended solids	Adults, breeding and brooding
Eutrophication	Adults, juveniles
Nitrogen	Adults, juveniles
Phosphate	Adults, juveniles
Dissolved oxygen	All stages
Conductivity	Adults, juveniles
Calcium	Adults, juveniles
pH (acidity)	Adults, juveniles
Interstitial particulates	*Juveniles*
Interstitial water chemistry	Juveniles
Industrial pollutants	All stages
Pesticides	All stages

Extensive mussel beds contribute to the health of rivers by their filtration power. Reductions in those beds reduces this ecological function. However replacement mollusc faunas have developed in some African reservoirs. River sections with steep gradients or escarpments sometimes offer optimal conditions for locating hydroelectric and other dams. However, those locations may provide special fastwater habitats for species with only scattered distribution, or local endemics. In some cases species may multiply following construction of a dam. The status of a freshwater pulmonate, Bulinus truneatus, changed from rare to common in Lake Volta, Ghana. This led to an increase in the level of urinary schistomiasis infections in the region.

In some tropical reservoirs the overall number of fish species has increased, although several riverine species have disappeared, e.g. Lakes Kariba, Zambia and Zimbabwe, and Ayame, Côte d' Ivoire,. Tilapias of the family Cichlidae are usually the most successful in these lakes. The extent of entrainment of larval and juvenile fishes in hydroelectric turbines varies according to flushing duration, depth of extraction and species present. Passage through turbines of young anadromous salmonids, en route downstream, is a well-known source of mortality. In the catadromous eels, family Anguillidae, it is the downstream migrating adults which are killed by the turbines.

Reservoirs promote waterfowl and many dams have substantial populations. The type of shoreline, shallow, with fringing vegetation, supports greater species diversity and larger numbers compared to steeply shelving mostly deep water sites. A substantial number of dammed sites support nationally or internationally important waterfowl assemblages. In more arid areas, creating dams increases numbers of birds able to remain all year round in otherwise largely seasonally dry places.

The presence of a dam has substantively altered the migration phenology and distribution of some species. Of 957 Ramsar sites 25% had natural lake types and 10% had artificial wetland types; 78% of the latter were dammed sites. Some of the Ramsar sites with dams supported internationally important waterfowl populations, though these sites were small in number compared to natural wetlands. The significance of such artificial wetlands for waterfowl is difficult to interpret. Nevertheless some dammed sites did prove suitable for large waterfowl assemblages.

Estuaries and coasts tend to have higher numbers of species because they attract waders as well as waterfowl, while inland systems generally support only wildfowl in internationally important numbers. A much smaller number of artificial wetlands are of international importance for wintering waterfowl. Only 11 support one or more internationally important populations and only three have large overall wintering populations >20,000 waterfowl. For several species a large proportion of nationally important lakes are artificial ones: 40% or more of important sites are artificial for eight waterfowl: little grebe, great crested grebe, great cormorant, gadwall, northern shoveller, pochard, tufted duck and common coot.

Much of South Africa is arid and has few natural permanent water bodies. Almost all permanent water bodies are dammed sites constructed for water storage. There are at least 517 major reservoirs and numerous small, farm dams. The presence of these new wetlands has had several consequences. At least 12 impoundments support major and important concentrations of waterfowl. Suitable conditions have been provided for the Pelecaniformes (pelicans, darters and cormorants), 70% of the global population of the South African shelduck during moulting, and refuges for species of national conservation concern such as the pink-backed pelican, Pelecanus rufescens, and the Caspian tern, Hydroprogne caspia.

Negative impacts in overall waterfowl assemblages in southern Africa are due to the loss of many of the former natural marshes and riverine habitats through reduction in river flow, removal of seasonal flow variability and consequent changes in sediment movement and channel stability. In addition poor dam management may cause sudden major releases of water, causing major downstream floods in areas that have had little or no flood activity for years. That can affect species that use unvegetated river banks and sand banks between river channels.

The construction of reservoirs converts lotic into lentic habitats. Species dependent on running water will diminish or disappear. In almost all cases, the diversity of fish species will drop. Reservoir fisheries are one of the frequently claimed benefits of impoundments. The changes in catches following impoundment are variable. However the catches in new reservoirs frequently go through a "boom and bust" cycle, with catches initially increasing following filling of the reservoir and then declining. Therefore impact assessments of dams should be based on the long term catches.

Water oscillations in the littoral zone reduce its suitability for species requiring stable conditions. Some mobile species, such as shorebirds, may find this habitat suitable for feeding. The nature of the deeper or profundal zone will depend on the climate, preparation of the reservoir prior to filling and other factors. In boreal and arctic areas the deeper waters are normally cooler than the surface waters and this provide habitats for both warm- and coolloving species. If the trees and other vegetation are not cleared from the reservoir, then decomposition commonly leads to low oxygen levels in the profundal zone usually only suitable for anoxic microorganisms.

The changed or fluctuating conditions in the reservoir may lead to opportunities for weed or exotic species e.g.

the water hyacinth, Eichhornia crassipe. Increases in the number of mollusc-borne diseases following dam construction in various countries.

Mercury is in a harmless inorganic form in many soils. Bacteria which feed on decomposing matter in a new reservoir transform the mercury into methyl mercury which passes up the food chain from plankton to fishes and those species that feed on fishes, including humans. Biomagnification results in higher mercury levels as mercury ascends the food chain. In the La Grande Phase of the James Bay, Canada, hydro project it was found that reservoir fish became contaminated with mercury at levels exceeding World Health Organisation (WHO) standards.

Sixtyfour percent of the Cree living in the La Grande estuary had blood mercury levels far exceeding WHO standards. Mercury can also negatively affect wildlife that prey on mercury-contaminated fishes. Loons are severely affected by mercury pollution. Sport fishing also plays a role in the local economy, and mercury contamination creates unfavourable publicity.

Reservoirs tend to serve as sediment traps since river velocities and carrying capacity for particles decrease in reservoirs. However sometimes fluctuating water levels in reservoirs erode the shores and add to the turbidity of the reservoir discharge. Sedimentation can degrade habitat both in the reservoir and below the dam, as well as reduce storage capacity. Many of the molluscan extinctions in the Mobile Bay, USA drainage, following multiple impoundments, were due to siltation. The degree of tolerance to silt cover depends on the species of mollusc.

Suspended silt may reduce the feeding efficiency of filter-feeding bivalves and other species. About 50 km^3 of sediment, nearly 1% of global reservoir capacity, was estimated in 1997 to be trapped behind dams. Keeping sediments flowing through reservoirs will benefit both reservoir life and the life of downstream ecosystems such as flood plains and deltas. Many of the existing potential solutions, such as dredging and sluicing, have economic or environmental limitations.

Downstream Segment

In the downstream segment, most of the impacts of a dam are negative. In a preliminary assessment of 66 case studies of the impact of dam construction on fishes, based on qualitative information, 73% of the impacts were negative and only 27% were positive. About 55% of the impacts were below the dam and linked to fish migrations and to floodplain access. In the distribution of molluscs along a 240 km stretch of the Little River in Oklahoma, USA there were mussel extinction gradients downstream from large impoundments. With increasing distance from the dam there was a relative increase in mussel species richness and species abundance. Only those stretches furthest from the dam contained the relatively rare species. Richness declined below each successive dam, with a multiplier effect.

Overall volume of flow: Some reservoirs have been filled by cutting off all or almost all flow downstream of the dam, e.g. Cabora Bassa, Mozambique with the consequent loss of organisms. Large aquatic species such as sturgeons, crocodiles and dolphins require minimal flows in which to navigate and feed. Such species may be affected by reduced flows including a reduction in the area of habitat utilised. This may lead to smaller populations, reduced growth rates and, where populations are already at risk, extirpation or extinction. A certain level of downstream flow is needed to maintain a minimum volume and area of habitat, oxygen concentration and other 'desirable' in-stream conditions and avoid lethal temperatures.

Normal seasonal flow patterns are a key to maintaining river biodiverstiy. Balancing reservoirs may help avoid pulse discharges, delay peak discharges and reduce them to an ecologically acceptable level and guarantee a certain minimum discharge. Constructing dams just above large tributaries can moderate changes to downstream flow patterns. If the tributary has a similar seasonal flow cycle to the mainstem, then the downstream flow pattern and seasonal cues will be less impacted than if the mainstem dam had been sited below the tributary.

Seasonal variability of flow and flood plains: The pattern of flow of a river undergoes a regular series of changes with the seasons. The patterns can differ profoundly from region to region e.g. in an Indian river the peak flow may be during the monsoon, in an Arctic river during snow-melt or ice break-up. The expansion and contraction of the river controls living space and access to particular habitats. It is to these profound seasonal patterns that the species in a drainage basin adapt. It is from river flow events that species take cues to migrate, spawn, etc.

Water table changes: Water diversion for irrigation may lower the downstream water table adjacent to the river. Further, the 'levelling out' of floods, may reduce seasonal recharging of the water table. The supply to springs, artesian flows and cave streams may also be affected. Cave streams and some aquifers sometime have species characterised by reduced or absent eyes, loss of pigmentation and enhanced non-visual sensory systems. According to David Culver there are 1,300 described obligatory cave species in the United States alone, while the undescribed species probably number several times that value. Many cave organisms are restricted to a single cave or cave complex.

Cutting off the cave water supply, either temporarily or permanently, may lead to extinctions. When selecting potential dam sites, planners should be alert to known cave dwelling speciesor check for the presence of unknown species. Of the world's 110 described species of cave fishes, a high proportion are threatened. Proudlove reviewed the conservation status of hypogean fishes. The percentage of threatened hypogean fish species has risen from 18% in 1977 to 87% in 1996. Most of the species have very limited distribution. The primary threats are habitat degradation (e.g. quarrying and siltation due to dams and logging), hydrological manipulations (e.g. water removal for human consumption or irrigation and damming), environmental pollution (e.g. eutrophication and poisoning from agricultural and industrial runoff), overexploitation (capture for the aquarium trade) and introduced exotic animals.

Abiotic changes: In summer, in temperate lakes, solar radiation heats the eplimnion but not the hypolimnion. The hypolimnion is often anoxic. In the autumn the lake undergoes mixing and some heat is transferred to the bottom `layers. Water discharge from the dam is usually below the epilimnion. Therefore in summer the water discharged into the river below the dam is colder and has less oxygen than normal and in winter it is warmer. These physical changes can effect the biota for long distances down the river.

Discharges from the reservoir are variable usually resulting from the requirements for hydroelectric power and not related to natural cycles. Flow below the dam can rapidly alter from almost standing water to torrential flows and water depth, water velocity, oxygen concentration, temperature, suspended solids, pollutants and chemical composition can change in a very short period of time. Many of the effects experienced downstream of a dam are in reverse of those produced in the reservoir above them. Heat, silt, inorganic and organic nutrients retained in the lake are lost to the stream below. Large annual variations in water level in the reservoir results in a decrease in the annual variation in water level in the efferent river.

Inland deltas: Dams trap sediments, diminishing the downstream supply. An inland delta and flood plain, including a network of oxbow lakes, in the middle Danube, was supplied by sediment during natural seasonal floods. The area produced an impressive harvest of trees, fishes and cereals. Biota included 65 species of fishes, 11 of amphibians, nine reptiles, 41 mammals and 242 birds. Construction of a series of dams, dredging the river channel and construction of a canal deprived the delta and flood plain of the annual supply of silt and resulted in severe alterations in the benthos and zooplankton communities as well as the change and decrease of species diversity and biomass.

The inland delta was lost eliminating spawning, feeding and overwintering grounds for fishes. Amongst fish species lost from the affected area were the percids, Zingel streber and Zingel zingel, both classed in the IUCN Red List as vulnerable, and the salmonids, Hucho hucho, classed by the IUCN Red List as endangered, and Salmo labrax m. fario. The European beaver, Castor fiber, has left the territories influenced by the dam. The mean annual fish catch has dropped by 87%.

Anti-gradient, thermal transport: North-flowing rivers such as the McKenzie River, Canada, transport warmer water into the Arctic than that from local tributaries. This enables some species to range further north. The same phenomenon would occur in reverse in the Southern Hemisphere. Rivers flowing towards the tropics or down from higher cooler altitudes would permit cool water fishes to extend their ranges.

Dams or series of dams may affect anti-gradient thermal transport and the ranges of aquatic species which depend on them. However the thermal transport affects more than just the species in the river. In the Arctic the northern limits of the tree-line, appears to be extend adjacent to north flowing rivers, as well as next to large lakes which act as reservoirs of summer heat. It is not clear whether reducing the flow of anti-gradient thermally transporting rivers, or the effects of reservoirs and discharge from either epilimnion or hypolimnion will affect the downstream and lateral distribution of species. Analysis of data from Russian rivers with older dams might be useful.

Water Basins

When waters of one basin are diverted into another, impacts can be expected from changes in volume and seasonality of flow. New biota from the source basin may invade the recipient basin and compete with the native species. If all the water is diverted from the source basin, this will obviously have serious impacts on any unique species or genetically different stocks. Individuals washed down irrigation canals, especially if there is a drop from the impoundment, may no longer be able to return and may not be able to maintain viable populations in the new habitat. In the Nicola River, western Canada, considerable numbers of Pacific salmon fry pass down into irrigation ditches, depleting the river populations.

Estuarine and Marine Impacts

Many of the effects in estuaries are similar to those upstream, e.g. loss of habitat and changes in seasonal flow, turbidity and productivity.

Estuarine deltas: Deltas below impoundments tend to shrink, reducing habitats, because of the capture of sediments by impoundments. The Nile Delta, Egypt, has shrunk at a rate of 125 to 175 m yr^{-1}, and more saline water has invaded inland. The Danube Delta, central Europe, shoreline is receding at a rate of up to 17 m yr^{-1}, threatening benefits from tourism to birdlife. The delta support large populations of bird species that are generally widespread over Europe; some 170 species of birds breed in the delta, including pelicans, herons, ibises and terns. Impoundments upstream, including 7 major dams, on the 2,860 km long river, channelisation and the loss of the nutrient absorption capacity of upstream floodlands has meant nutrients and other pollutants are affecting delta water quality. Bird populations are at a fraction of their historical numbers. So although reservoirs may provide new habitat upstream their impacts on birds may be negative in the long-term.

Salinity, nutrients and reproduction: Decreased discharge rates can result in an increase in salinity in estuaries and change the composition of species in this zone. The effects of increased salinity on fishes of the Nile Delta, Egypt, has been documented by several authors. Abramovitch, for example notes that out of 47 commercial fish species in the Nile prior to the construction of the Aswan High Dam, only 17 were still harvested a decade after its completion. The annual sardine harvest in the eastern Mediterranean has dropped by 83%, probably the effect of a reduction in nutrient-rich silt entering that part of the sea. The effect of lowered nutrient input is generally the greatest in the first year of life of the fishes. Rivers can increase nutrient levels at river mouths by two processes: entrainment and transport.

Non-dam Impacts

Dams often interact with the effects of other human activities. McAllister classes the potential interaction of human impacts as either neutralising, additive or synergistic He postulated that synergistic interactions, those where harmful affects were amplified, were more likely, considering that genomes by selection are adapted to a certain set of undisturbed natural environmental conditions. Species classed as endangered are more likely to become extinct. Factors to be considered include climate change agriculture, forestry, industry and municipal effects.

The exception to this generalisation is the migratory species that move up and downstream and use such movements to maintain genetic diversity or the survival of a species in that part of the basin, or in their entire global range. For molluscs and fishes the change from running to still waters in a newly established reservoir is profound. Most species highly adapted to currents will be extirpated and the reservoir diversity will drop. The reservoir becomes stocked with ecologically flexible native species or with exotics. One exception is in African reservoirs where fish numbers may increase.

Reservoir fishery harvests usually increase after impoundment, but then drop. For waterfowl the situation is different. Reservoirs provide new habitat for over-wintering in cool regions and for residence in warm arid regions which have few natural water bodies. Thus new reservoirs can increase waterfowl populations, though their diversity will not be as high as that in natural lakes. The effects of impoundment on birds that existed in the reservoir basin and downstream of the dam have not been well studied.

The impounding of rivers has terrestrial impacts on biodiversity. The biodiversity of land flooded by reservoirs, and floodplains, wetlands, oxbows and other river valley aquatic ecosystems deprived of normal flooding may be diminished or lost. River ecology is tied to that of estuaries in the transport of silt, nutrients and seasonally different volumes of river discharge. This is important in the physical maintenance of delta and coastal habitats and the nutrient-based estuarine food chains. The nutrient plume of rivers can extend far out to sea. So regulation of rivers can influence even ocean species and ecosystems.

Cumulative Impacts

The addition of each new dam in a river contributes to the fragmentation of habitat and separation of populations. Gene flow, hitherto bidirectional, becomes unidirectional, downstream, reducing genetic diversity. Each new dam also prevents natural restoration of upstream populations lost through natural or anthropogenic causes. One of the biggest cumulative impacts may be that a greater proportion of running water is converted to still reservoirs habitat.

Standards for Minimising Negative Impacts on Biodiversity

As a result of the continuing erosion of biological diversity, the inter-governmental community has, through a variety of mechanisms, adopted a set of standards for minimising harmful impacts on biodiversity. These standards, which have been adopted in legally binding documents by almost all governments, are far more exacting and demanding than is generally recognised.

World Charter for Nature

The World Charter for Nature was adopted by consensus by the UN General Assembly in 1982. It provides the high-level guiding principles that should govern human responsibility for biodiversity. Its states that activities which might have an impact on nature shall be controlled, and the best available technologies that minimise significant risks to nature or other adverse effects shall be used; in particular:

— Activities which are likely to cause irreversible damage to nature should be avoided;

— Activities which are likely to pose a significant risk to nature shall be preceded by an exhaustive examination; their proponents shall demonstrate that expected benefits outweigh potential damage to nature, and where potential adverse effects are not fully understood, the activities should not proceed;

— Activities which may disturb nature shall be preceded by assessment of their consequences, and environmental impact studies of development projects shall be constructed in advance, and if they are to be undertaken, such activities shall be planned and carried out so as to minimise potential adverse impacts.

Convention on Biological Diversity

The Convention on Biological Diversity (CBD) was signed by 156 States in June 1992, and by September 1999 175 countries had ratified the Convention. The Preamble to the CBD starts with similar sentiments to those in the World Charter for Nature. The objectives of the CBD are the conservation of biological diversity, the sustainable use of its components, and the fair and equitable sharing of the benefits arising out of the utilisation of genetic resources. Article 8 of the CBD identifies the in-situ conservation measures that Parties should seek to carry out. Article 14 addresses impact assessment and minimising adverse impacts, inter alia in Section 1 (a): 'Introduce appropriate procedures requiring environmental impact assessment of its proposed projects that are likely to have significant adverse effects on biological diversity with a view to avoiding or minimising such effects and, where appropriate, allow for public participation in such procedures'.

Other Agreements

A very similar set of standards relating to impacts on biodiversity is included in Agenda 21, which was adopted by consensus at the UN Conference on Environment and Development in June 1992. Chapter 15 on Conservation of Biological Diversity states that: processes and activities with significant impacts on biological diversity should be identified; action should be taken for the conservation of biological diversity through the in-situ conservation of ecosystems and natural habitats; and that the rehabilitation and restoration of damaged ecosystems and the recovery of threatened and endangered species should be promoted. The Convention on the Conservation of Migratory Species of Wild Animals (CMS), adopted in 1979, states, among its fundamental principles, that: 'the Parties acknowledge the need to take action to avoid any migratory species becoming endangered'. This principle is of particular importance for the current report, since migratory freshwater species have suffered particularly serious adverse affects from large dams..

At the national level, numerous laws, action plans, and conservation programmes have enshrined the principles contained in the documents quoted above.

World Bank

The World Bank's Operational Policy 4.04 on Natural Habitats requires that comprehensive analysis should demonstrate that overall benefits from a project outweigh the environmental costs before significant conversion of natural habitats is allowed, unless there are no feasible alternatives for the project and its siting. The World Bank Environment Department study, the Impact of Environmental Assessment reported some positive trends but also a wide range of quality in the biodiversity sections of Environmental Impact Assessment (EIA) reports

Corporate Sector

The World Business Council for Sustainable Development and the IUCN have prepared: Business and biodiversity. A guide for the private sector. This explains the Biodiversity Convention, issues and opportunities for the private sector, and describes how companies can engage with biodiversity issues. Hydro-Québec, Canada, a major hydro-power company, is moving into wind-generated power. The company has co-published a report on biodiversity and hydroelectricity.

Government Sector

Some governments have high environmental protection standards. The United States Environmental Protection Act was a pioneer in establishing standards not in outlawing killing endangered species but in protecting their habitat. Work on a California dam in 1996 included substantive pro-active environmental activities. Endangered frogs were relocated. Flooded habitat had to be replaced and artificial ponds which would be self sustaining were created with large reed beds for amphibians and birds. There were stiff penalties for any construction worker moving outside the terms and conditions of site work, including loss of job and financial penalties. Training videos were produced which even visitors to the site had to watch.

Environmental Impact Assessment Standards

UNEP and Wetlands International have been promoting economic and environmental impact assessment. Dorcey et al. provide a checklist for key potential environmental and social impacts caused by large dam projects. The public expects high standards for environmental impact surveys (EIA), an expectation not always met. Standard most frequently breached are:

Lead time: EIAs should begin long before any construction, especially in geographic regions where biota are poorly known. Lead time should permit sampling through at least one annual cycle, preferably more to allow for natural variability. Biodiversity monitoring should be continued through the construction phase.

Expertise: Taxonomic knowledge is required. Similar expertise is required in ecological, life history, fishery and other aspects. It is desirable to complement academic expertise with local indigenous and traditional knowledge.

Open process: Draft and final EIA studies should be freely available locally, nationally and internationally with sufficient lead-time before public hearings and final approval processes. Voucher specimens. Representative voucher specimens of flora and fauna collected should be deposited in national or regional museums.

'Arm's length' analysis: EIA's should not be conducted by the firms engaged in building the dam, nor their subsidiaries.

Follow-up: One-, five- and ten-year follow-up biodiversity studies, by an organisation other than the one which carried out the EIA, should be performed. These test EIA predictions and provide data for planning future dams. EIAs should consider what are alternative uses of the site, in addition to current usage and the planned dam. What are the costs and benefits of the various options, including items that do not enter the normal market such as subsistence fishing? What is the lifetime of the various options? In some cases dam capacity has been significantly reduced in less than 10 years, in other cases dams are functioning at good capacity several decades after their construction. Can sedimentation be mitigated by engineering solutions in situ or reduced ex situ by biological management of the drainage basin? What are the biodiversity effects of these various options? If heavy sediment deposition is impossible to avoid, is this a factor in possible elimination of this dam site?

10

Endangered Species

An endangered species is a species whose population is so small that it is in danger of becoming extinct. Its numbers are usually low, and it needs protection in order to survive. According to scientists, more than one and one-half million species exist on the earth today. There are more than 1,000 animal and plant species endangered worldwide. Aquatic species, which are often overlooked, are facing serious trouble.

Many countries have laws offering special protection to these species to prevent their extinction. There are more than 3,500 protected areas in existence worldwide. These areas include parks, wildlife refuges and other reserves. They cover a total of nearly 2 million square miles, or 3% of our total land area. Only a few of the many truly endangered species actually make it to the lists and obtain legal protection.

Many more species become extinct, or potentially will become extinct, without gaining public notice. Many of these laws are controversial. Typical areas of controversy include: criteria for placing a species on the endangered species list, and criteria for removing a species from the list once its population has recovered; whether restrictions on land development constitute a "taking" of land by the government, and the related question of whether private landowners should be compensated for the loss of use of their land; and obtaining reasonable exceptions to protection laws.

Listing of Endangered Species

All most all countries maintain a list of endangered and threatened species. Any person may petition the government to list a species as either endangered or threatened. A declining species has to be added to the official list of endangered and threatened species before it receives any protection. But just getting on the list can be the hardest part.

A listing as an endangered species can backfire, as it makes a species more valuable and more desirable for collectors and poachers. The conservation status of a species is an indicator of the likelihood of that endangered species continuing to survive. Many factors are taken into account when assessing the conservation status of a species; not simply the number remaining, but the overall increase or decrease in the population over time, breeding success rates, known threats, and so on.

The best-known worldwide conservation status listing is the IUCN Red List, but many more specialised lists exist. The following conservation status categories are loosely based on the IUCN categories.

— *Extinct:* the last remaining member of the species had died, or is presumed to have died beyond reasonable doubt. Examples: Thylacine, Dodo.

— *Extinct in the wild:* captive individuals survive, but there is no free-living, natural population. Examples: Dromedary, Przewalski's Horse.

— *Critical or critically endangered:* faces an extremely high risk of extinction in the immediate future. Examples: Ivory-billed Woodpecker, Arakan Forest Turtle

— *Endangered:* faces a very high risk of extinction in the near future. Examples: Kings holly, Pink fairy armadillo

— *Vulnerable:* faces a high risk of extinction in the medium-term. Examples: Cheetah, Bactrian Camel

— *Secure or lower risk:* no immediate threat to the survival of the species. Examples: Nootka Cypress, Llama

List of Endangered Animals and Plants

Details of some of the important endangered species are described in this section. Note that because of varying standards for regarding a species as endangered, and the very large number of endangered species, the given list should not be regarded as comprehensive.

Endangered Mammals

Elephants

— *African Elephant (Loxodonta africana):* The african elephant is classified as a member of the order *Proboscidea* (*Elephants*) and is a member of the family *Elephantidae*. The male stands up to ten feet high to its shoulder, and weighs up to six tons. The female is slightly smaller, and weighs up to four tons. It is classified as an endangered species due to a reduction of at least 50% of the african elephant population over the last three generations based on an index of abundance. Hunting of the african elephant is now banned in several countries, but poaching for ivory still exists.

— *Asian Elephant (Elephas maximus)*: The Asian elephant is classified as a member of the order *Proboscidea* (*Elephants*) and is a member of the family Elephantidae. It stands up to ten feet high and twenty feet long. It weighs up to 10,000 pounds. The asian elephant is classified as an endangered species due to a reduction of at least 50% of the asian elephant population over the last three generations based on an index of abundance and a decline in area of occupancy. The Asian elephant has four subspecies: the Indian, Ceylon, Sumatran, and Malaysian elephants.

Whales

— *Right Whale (Eubalaena glacialis)*: The right whale is classified as a member of the order *Cetacea* (*Whales*) and is a member of the family Balaenidae. It grows up to sixty feet long, and is twelve to eighteen feet long at birth. It weighs up to sixty tons as an adult. The right whale is classified as an endangered species due to an estimated population of less than 250 mature right whales and an estimated continuing decline of at least 20% within two generations. The right whale was once the most hunted of all whales, and is now protected by law.

— *Blue Whale (Balaenoptera musculus)*: The blue whale is classified as a member of the order *Cetacea* (*Whales*) and is a member of the family Balaenopteridae. The male blue whale grows to about eighty-two feet long, and the female grows to about eighty-five feet long. It weighs up to 285,000 pounds as an adult. The blue whale is classified as an endangered species due to a reduction of at least 50% of the blue whale population over the last three generations based on direct observation, an index of abundance, and actual levels of exploitation. The blue whale is the largest mammal to have lived on the earth, but it feeds on some of the smallest marine organisms: plankton.

— *Fin Whale (Balaenoptera physalus)*: The fin whale is classified as a member of the order *Cetacea* (*Whales*) and is a member of the family Balaenopteridae. It grows up to eighty feet long. The fin whale is classified as an endangered species due to a reduction of at least 50% of the blue whale population over the last three generations based on direct observation, an index of abundance, and actual levels of exploitation.

Primates

— *Golden Lion Tamarin (Leontopithecus rosalia)*: The golden lion tamarin is classified as a member of the order Primates and is a member of the family Callitrichidae. The golden lion tamarin's head and body are about one foot long, and the tail is slightly shorter. It weighs about one and a half pounds. The golden lion tamarin is classified as a critically endangered species due to the fact that it only exists in severely fragmented subpopulations consisting of no more than fifty mature tamarins each, and that there has been continuing decline in the golden lion tamarin population. The golden lion tamarin is one of the most endangered of all mammals.

— *Hybrid Spider Monkey (Ateles belzebuth hybridus)*: The hybrid spider monkey is classified as a member of the order Primates and is a member of the family Cebidae. The hybrid spider monkey grows to almost two feet long, not including the tail. It weighs from ten to fifteen pounds. The hybrid spider monkey is classified as an endangered species due to the fact that it only exists in severely fragmented subpopulations, and that there has been continuing decline in the hybrid spider monkey population. The hybrid spider monkey is known for its ability to use its tail as an extra limb.

— *Aye-aye (Daubentonia madagascariensis)*: The aye-aye is classified as a member of the order Primates and is a member of the family *Daubentoniidae*. The aye-aye is about the size of a rabbit, and is brown. It is a nocturnal animal. The aye-aye is classified as an endangered species due to a projected reduction of at least 50% of the aye-aye population over the next ten years based on levels of exploitation and a decline in area of occupancy. Also, the aye-aye has an estimated population of less than 2500 and an observed continuing decline in the form of severly fragmented subpopulations. The aye-aye builds nests out of twigs to hide during the day. It can be found on the African island of Madagascar.

— *Gorilla (Gorilla gorilla)*: The gorilla is classified as a member of the order Primates and is a member of the family Hominidae. The male gorilla grows to about six feet high, and weighs up to six hundred pounds. The female gorilla grows to about five feet high, and weighs up to two hundred pounds. The gorilla is classified as an endangered species due to the projected gorilla population declining to at the highest 50% due to a decline in area of occupancy. The gorilla is the largest and most powerful primate alive, but is a peaceful and sociable animal.

Carnivores

— *Red Wolf (Canis rufus)*: The red wolf is classified as a member of the order *Carnivora* (*Carnivores*) and is a member of the family Canidae. The red wolf is classified as a critically endangered species due to the estmation that its population consists of less than fifty mature red wolves.

— *Amur Leopard (Panthera pardus orientalis):* The amur leopard is classified as a member of the order Carnivora (Carnivores) and is a member of the family Felidae. It is classified as a critically endangered species due to a reduction of at least 80% of its population over the past three generations because of a decline in area of occupancy, extent of occurance, or quality of habitat. The amur leopard population is also estimated to be less than 50 mature individuals. Amur leopards can be found in eastern Asia.

— *Anatolian Leopard (Panthera pardus tulliana)*: The anatolian leopard is classified as a member of the order *Carnivora* (*Carnivores*) and is a member of the family Felidae. It is classified as a critically endangered species due to an estimated population of less than 250 mature individuals and a continuing decline in numbers of mature individuals and population structure in the form of severely fragmented populations. Anatolian leopards can be found in Turkey.

— *Asiatic Cheetah (Acinonyx jubatus venaticus)*: The Asiatic cheetah is classified as a member of the order *Carnivora* (*Carnivores*) and is a member of the family Felidae. It is classified as a critically endangered species due to an estimated population of less than 50 mature individuals and a continuing decline in numbers of mature individuals and population structure due to the fact that all Asiatic cheetahs are in a single population. Asiatic cheetahs can be found in Iran.

— *Florida Cougar (Puma concolor coryi)*: The Florida cougar is classified as a member of the order *Carnivora* (*Carnivores*) and is a member of the family Felidae. It is also known as the Florida panther and the Florida

Puma. It is classified as a critically endangered species due to an estimated population of less than 50 mature individuals. Florida cougars can be found in the United States.

— *Iberian Lynx (Lynx Pardinus)*: The iberian lynx is classified as a member of the order *Carnivora* (*Carnivores*) and is a member of the family Felidae. It is classified as an endangered species due to an estimated population of less than 2500 mature individuals and an extimated continuing decline of at least 20% of its population within two generations. Iberian lynx can be found in Portugal and Spain.

— *Snow Leopard (Uncia uncia)*: The snow leopard is classified as a member of the order *Carnivora* (*Carnivores*) and is a member of the family Felidae. It is classified as an endangered species due to an estimated population of no more than 2500 snow leopards and the fact that it has no subpopulation numbering more than 250 mature leopards. Snow leopards can be found in eastern Asia.

— *Texas Ocelot (Leopardus pardalis albescens)*: The Texas ocelot is classified as a member of the order *Carnivora* (*Carnivores*) and is a member of the family Felidae. It is classified as an endangered species due to an estimated population of less than 250 mature individuals. The Texas ocelot can be found in Mexico and the United States.

— *Tiger (Panthera tigris)*: The tiger is classified as a member of the order *Carnivora* (Carnivores) and is a member of the family *Felidae*. The male tiger grows up to ten feet long from its head to the tip of its tail, and weighs up to 575 pounds. The tiger is classified as an endangered species due to the projected tiger population declining to at the highest 50% due to an index of abundance and a decline in area of occupancy. The tiger consists of eight subspecies, distinguished by the colour of their coat.

— *Marine Otter (Lutra felina)*: The marine otter is classified as a member of the order *Carnivora* (*Carnivores*) and is a member of the family Mustelidae. It is classified as an endangered species due to the marine otter population declining because of the levels of exploitation and a decline in its area of occupancy. Marine otters can be found in South America.

— *Giant Panda (Ailuropoda melanoleuca)*: The giant panda is classified as a member of the order *Carnivora* (*Carnivores*) and is a member of the family Procyonidae. The male giant panda stands up to five feet tall, and weighs up to 265 pounds. The female giant panda is smaller and weighs less. The giant panda is classified as an endangered species due to the fact that it only exists in severely fragmented subpopulations consisting of up to 250 mature adults, and that there has been continuing decline in the area of habitat. The giant panda feeds mainly on bamboo, even though it is classified as a carnivore.

— *Lesser Panda (Ailurus fulgens):* The lesser panda is classified as a member of the order *Carnivora* (*Carnivores*) and is a member of the family Ursidae. It is classified as an endangered species due to an estimated population of less than 2500 mature individuals and a continuing decline in numbers of mature individuals because of severly fragmented populations. Lesser pandas can be found in southern Asia.

Even-toed hoofed mammals

— *Cuvier's Gazelle (Gazella cuvieri)*: The Cuvier's gazelle is classified as a member of the order *Artiodactyla*

(*Even-Toed Hoofed Mammals*) and is a member of the family Bovidae. It is classified as an endangered species due to a severely fragmented population estimated to number less than 2500 mature individuals. The Cuvier's gazelle can be found in Algeria, Morocco, and Tunisia.

— *Western Giant Eland (Tragelaphus derbianus derbianus)*: The western giant eland is classified as a member of the order *Artiodactyla* (*Even-Toed Hoofed Mammals*) and is a member of the family Bovidae. The western giant eland is one of the largest antelopes in Africa. It can stand up to six feet high at the shoulder. It weighs up to 1,500 pounds. The western giant eland is classified as an endangered species due to a population estimated to number less than 2500 mature individuals, an estimated continuing decline of at least 20% within 2 generations, and a continuing decline observed by the fact that all individuals are in a single subpopulation. The eland has long spiraled horns and tufted, cattlelike tails. It can be found in western Africa.

— *Wild Bactrian Camel (Camelus bactrianus)*: The wild Bactrian camel is classified as a member of the order *Artiodactyla* (*Even-Toed Hoofed Mammals*) and is a member of the family Camelidae. The wild Bactrian camel's shoulder height is about six feet, and it grows up to twelve feet long. It weighs up to 1,500 pounds. The wild Bactrian camel is classified as an endangered species due to a reduction of at least 50% of the wild Bactrian camel population over the last three generations based on a decline in area of occupancy and actual levels of exploitation, and due to severely fragmented subpopulations numbering no more than 250 mature camels. The wild Bactrian camel is the only truly wild two-humped camel in the world.

— *Manipur Brow-Antlered Deer (Cervus eldi eldi)*: The Manipur brow-antlered deer is classified as a member of the order *Artiodactyla (Even-Toed Hoofed Mammals)* and is a member of the family Cervidae. It is classified as a critically endangered species due to a continuing decline in area of occupancy and the fact that there are no more than 250 mature Manipur brow-antlered deers and they live in a single population. These animals can be found in India.

Odd-toed hoofed mammals

— *Black Rhinoceros (Diceros bicornis)*: The black rhinoceros is classified as a member of the order *Perissodactyla (Odd-Toed Hoofed Mammals)* and is a member of the family *Rhinocerotidae.* The black rhinoceros grows up to twelve feet long and six feet high. It weighs up to three thousand pounds. The black rhinoceros is classified as a critically endangered species due to a reduction of at least 80% of the black rhinoceros population over the last three generations based on direct observation, an index of abundance, and a decline in area of occupancy. The black rhinoceros is the most aggressive species in the rhinoceros family, and can charge at speeds up to thirty miles per hour.

Marsupials

— *Broom's Pygmy-possum (Burramys parvus)*: The Broom's Pygmy-possum is classified as a member of the order Marsupialia (Marsupials) and is a member of the family Burramyidae. It is also known as the mountain pygmy-possum. The Broom's pygmy-possum is classified as an endangered species due to an estimated extent of occurrence that is less than 5000 km^2, a severely fragmented population, and a projected decline in area of occupancy, extent of occurance, quality of habitat, number of subpopulations, and number of mature individuals. The Broom's pygmy-possum can be found in Australia.

— *Northern Hairy-nosed Wombat (Lasiorhinus krefftii)*: The northern hairy-nosed wombat is classified as a member of the order *Marsupialia* (*Marsupials*) and is a member of the family Vombatidae. It is classified as a critically endangered species due to an estimated extent of occurrence that is less than 100 km², a severely fragmented population, a projected decline in quality of habitat, and an estimated population of less than 50 mature individuals. The northern hairy-nosed wombat can be found in Australia.

Rodents

— *Short-tailed Chinchilla (Chinchilla brevicaudata)*: The short-tailed chinchilla is classified as a member of the order Rodentia (Rodents) and is a member of the family Chinchillidae. It is classified as a critically endangered species due to an estimated population decline of at least 80% over the past ten years based on a decline in area of occupancy and levels of exploitation. The short-tailed chinchilla can be found in South America.

Edentates

— *Giant Armadillo (Priodontes maximus)*: The giant armadillo is classified as a member of the order *Edentata* (*Edentates*) and is a member of the family Dasypodidae. It is classified as an endangered species due to an estimated population decline of at least 50% over the past ten years based on a decline in area of occupancy and levels of exploitation. The giant armadillo can be found in South America.

Endangered Birds

Screamers and duckss

— *Crested Shelduck (Tadorna cristata)*: The crested shelduck is classified as a member of the order Anseriformes (Screamers and ducks) and is a member of the family Anatidae. It is classified as a critically endangered species due to an estimated population of less than 50 mature crested shelducks. The species can be found in eastern Asia.

— *White-winged Duck (Cairina scutulata)*: The white-winged duck is classified as a member of the order Anseriformes (Screamers and ducks) and is a member of the family Anatidae. It is classified as an endangered species due to an estimated population of less than 2500 mature white-winged ducks, an estimated continuing decline of the population, and a severely fragmented population. The white-winged duck can be found in southern Asia.

Kingfishers, hornbills, and allies

— *Marquesan Kingfisher (Todirhamphus godeffroyi)*: The marquesan kingfisher is classified as a member of the order Coraciiformes (Kingfishers, hornbills, and allies) and is a member of the family Alcedinidae. It is classified as an endangered species due to a severely fragmented population and a projected decline of mature marquesan kingfishers.

— *Rufous-lored Kingfisher (Todirhamphus winchelli)*: The rufous-lored kingfisher is classified as a member of the order Coraciiformes (Kingfishers, hornbills, and allies) and is a member of the family Alcedinidae. The rufous-lored kingfisher is classified as an endangered species due to a reduction of at least 50% of the rufous-lored kingfisher population over the last three generations based on a decline in area of occupancy, and a projected decline of at least 50% over the next three generations based on the same reason.

Eagles, hawks, and vultures

— *California Condor (Gymnogyps californianus)*: The California condor is classified as a member of the order *Falconiformes* (*Eagles, hawks, and vultures*) and is a member of the family Cathartidae. It is classified as aa critically endangered species due to an estimated population of less than 50 mature individuals. The California condor can be found in the United States.

— *Mauritius Kestrel (Falco punctatus)*: The Mauritius kestrel is classified as a member of the order *Falconiformes* (*Eagles, hawks, and vultures*) and is a member of the family Falconidae. The Mauritius kestrel is classified as an endangered species due to an estimated population of less than 250 mature Mauritius kestrels.

Passerines

— *Hawaiian Crow (Corvus hawaiiensis)*: The Hawaiian crow is classified as a member of the order *Passeriformes* (*Passerines*) and is a member of the family Corvidae. The Hawaiian crow is classified as a critically endangered species due to an estimated population of less than 50 mature Hawaiian crows.

— *Cochabamba Mountain-finch (Poospiza garleppi)*: The cochabamba mountain-finch is classified as a member of the order Passeriformes (Passerines) and is a member of the family Emberizidae. The cochabamba mountain-finch is classified as an endangered species due to the existence of only a single population numbering no more than 2500 mature adults, a decline in area of habitat, and an estimated decline of at least 20% over the next ten years.

— *Gouldian Finch (Chloebia gouldiae)*: The gouldian finch is classified as a member of the order *Passeriformes* (*Passerines*) and is a member of the family Estrildidae. The gouldian finch is classified as an endangered species due to the fact that it only exists in severely fragmented subpopulations, and that a continuing decline is projected in the number of subpopulations and the number of mature adults.

Parrots

— *Blue-bellied Parrot (Tridaria malachitacea):* The blue-bellied parrot is classified as a member of the order *Psittaciformes* (*Parrots*) and is a member of the family Psittacidae. The blue-bellied parrot is classified as an endangered species due to the fact that it only exists in severely fragmented subpopulations of no more than 250 mature parrots each.

— *Whooping Crane (Grus americana):* The whooping crane is classified as a member of the order *Gruiformes* (*Cranes, rails, trumpeters*) and is a member of the family Gruidae. It is classified as an endangered species due to an estimated population of less than 250 mature cranes. The whooping crane can be found in Canada and the United States.

Endangered Fish

Acipenseriformes

— *Siberian Sturgeon (Acipenser baerii baerii)*: The Siberian sturgeon is classified as a member of the order *Acipenseriformes* and is a member of the family Acipenseridae. It is classified as an endangered species due to a projected reduction of at least 50% of the

Siberian sturgeon population over the next ten years based on actual levels of exploitation.

— *Alabama Sturgeon (Scaphirhynchus suttkusi)*: The Alabama sturgeon is classified as a member of the order *Acipenseriformes* and is a member of the family Acipenseridae. It is classified as a critically endangered species due to a reduction of at least 80% of the Alabama sturgeon population over the last ten years based on a decline in area of occupancy and actual levels of exploitation, and a projected decline of at least 80% over the next ten years based on the same reason.

— *Chinese Paddlefish (Psephurus gladius)*: The Chinese paddlefish is classified as a member of the order *Acipenseriformes* and is a member of the family Polyodontidae. It is classified as a critically endangered species due to a projected reduction of at least 80% of the Alabama sturgeon population over the next ten years based on a decline in area of occupancy and actual levels of exploitation.

Clupeiformes (Herrings)

— *Alabama Shad (Alosa alabamae)*: The Alabama shad is classified as a member of the order *Clupeiformes (Herrings)* and is a member of the family Clupeidae. It is classified as an endangered species due to a severely fragmented population and a continuing decline in the number of mature adults.

Cypriniformes

— *Silver Shark (Balantiocheilos melanopterus)*: The silver shark is classified as a member of the order Cypriniformes and is a member of the family Cyprinidae. The silver shark is classified as an endangered species due to a reduction of at least e0%

of the silver shark population over the last ten years based on direct observation and a decline in area of occupancy.

— *Clanwilliam Redfin (Barbus calidus)*: The clanwilliam redfin is classified as a member of the order Cypriniformes and is a member of the family Cyprinidae. The clanwilliam redfin is classified as an endangered species due to the fact that its estimated area of occurence is less than 5000 km^2, it is known to exist at no more than five locations, and continuing decline is projected for extent of occurence and area of habitat.

— *Wild Common Carp (Cyprinus carpio)*: The wild common carp is classified as a member of the order Cypriniformes and is a member of the family Cyprinidae. It is classified as a critically endangered species due to a projected reduction of at least 80% of the wild common carp population over the next ten years based on a decline in area of occupancy and introduced pollutants, parasites, or other threats.

Endangered Plants

About 6% of the 300,000 identified species are endangered due to overcollection or destruction of habitat, among other causes. Pollinator decline is also a factor for some species.

— African violet (*Saintpaulia ionantha*), due to forest clearance.

— Baishanzu fir (*Abies beshanzuensis*) of southeast China, three trees known on an isolated mountain summit.

— Baker's larkspur (*Delphinium bakeri*) of California, due to very limited habitat.

— Chilean wine palm (*Juba chilensis*), due to land clearance.

— Dawn Redwood ("*Metasequoia glyptostroboides*"), thought to be extinct until 1941, when a small stand was discovered in China.

— King of the Paphs Orchid (*Paphiopedilum rothschildianum*) of Asia, due to overcollection.

— Lobster claw (*Clianthus puniceus*) of Australia, due to overgrazing.

— Louisiana Quillwort, (*Isoetes louisianensis*) of Louisiana, due to very limited habitat.

— Madonna lily (*Lilium candidum*) of Europe, due to overcollection.

— Pinus squamata of southwest China, about 20 trees known.

— Saguaro cactus (*Carnegia gigantea*) of North America, due to overcollection, slow maturing, and slow breeding.

— Saharan Cypress (*Cupressus dupreziana*) of North Africa, due to small population and desertification.

— Wollemi Pine (*Wollemia nobilis)* of Australia, also known as the 'Dinosaur Tree' or 'Living Fossil'. Wollemia fossils have been found in Australia, Antarctica and New Zealand. The plant was thought to be extinct until two trees were discovered in 1994. Research into the horticultural development of the Wollemi pine is being conducted at Mount Annan Botanic Garden and, with commercial propagation well under way, plants should be available in 2005.

Causes of Endangerment

There are many reasons why a particular species may become endangered. Although these factors can be analysed and grouped, there are many causes that appear repeatedly. Below are several factors leading to endangerment:

Destruction of Habitat

Our planet is continually changing, causing habitats to be altered and modified. Natural changes tend to occur at a gradual pace, usually causing only a slight impact on individual species. However, when changes occur at a fast pace, there is little or no time for individual species to react and adjust to new circumstances. This can create disastrous results, and for this reason, rapid habitat loss is the primary cause of species endangerment.

The strongest forces in rapid habitat loss are human beings. Nearly every region of the earth has been affected by human activity, particularly during this past century. The loss of microbes in soils that formerly supported tropical forests, the extinction of fish and various aquatic species in polluted habitats, and changes in global climate brought about by the release of greenhouse gases are all results of human activity.

It can be difficult for an individual to recognise the effects that humans have had on specific species. It is hard to identify or predict human effects on individual species and habitats, especially during a human lifetime. But it is quite apparent that human activity has greatly contributed to species endangerment. For example, although tropical forests may look as though they are lush, they are actually highly susceptible to destruction. This is because the soils in which they grow are lacking in nutrients. It may take Centuries to re-grow a forest that was cut down by humans or destroyed by fire, and many of the world's severely threatened animals and plants live in these forests. If the current rate of forest loss continues, huge quantities of plant and animal species will disappear.

Introduction of Exotic Species

Native species are those plants and animals that are part of a specific geographic area, and have ordinarily been a part of that particular biological landscape for a lengthy period of time. They are well adapted to their local environment and are accustomed to the presence of other native species within the same general habitat.

Exotic species, however, are interlopers. These species are introduced into new environments by way of human activities, either intentionally or accidentally. These interlopers are viewed by the native species as foreign elements. They may cause no obvious problems and may eventual be considered as natural as any native species in the habitat. However, exotic species may also seriously disrupt delicate ecological balances and may produce a plethora of unintended yet harmful consequences.

The worst of these unintended yet harmful consequences arise when introduced exotic species put native species in jeopardy by preying on them. This can alter the natural habitat and can cause a greater competition for food. Species have been biologically introduced to environments all over the world, and the most destructive effects have occurred on islands. Introduced insects, rats, pigs, cats, and other foreign species have actually caused the endangerment and extinction of hundreds of species during the past five centuries. Exotic species are certainly a factor leading to endangerment.

Over-exploitation

A species that faces over-exploitation is one that may become severely endangered or even extinct due to the rate in which the species is being used. Unrestricted whaling during the 20th century is an example of overexploitation, and the whaling industry brought many species of whales to extremely low population sizes.

When several whale species were nearly extinct, a number of nations agreed to abide by an international moratorium on whaling. Due to this moratorium, some whale species, such as the grey whale, have made remarkable comebacks, while others remain threatened or endangered. Due to the trade in animal parts, many species continue to suffer high rates of exploitation. Even today, there are demands for items such as rhino horns and tiger bones in several areas of Asia. It is here that there exists a strong market for traditional medicines made from these animal parts.

Other Factors

Disease, pollution, and limited distribution are more factors that threaten various plant and animal species. If a species does not have the natural genetic protection against particular pathogens, an introduced disease can have severe effects on that specie. For example, rabies and canine distemper viruses are presently destroying carnivore populations in East Africa.

Domestic animals often transmit the diseases that affect wild populations, demonstrating again how human activities lie at the root of most causes of endangerment. Pollution has seriously affected multiple terrestrial and aquatic species, and limited distributions are frequently a consequence of other threats; populations confined to few small areas due to of habitat loss, for example, may be disastrously affected by random factors.

Importance of Protecting the Endangered Species

Plants and animals hold medicinal, agricultural, ecological, commercial and aesthetic/recreational value. Endangered species must be protected and saved so that future generations can experience their presence and value.

Medicinal Value

Plants and animals are responsible for a variety of useful medications. In fact, about forty percent of all prescriptions written today are composed from the natural compounds of different species. These species not only save lives, but they contribute to the growth of pharmaceutical industry.

Unfortunately, only 5% of known plant species have been screened for their medicinal values, although we continue to lose up to 100 species daily. The Pacific yew, a slow-growing tree found in the ancient forests of the Pacific Northwest, was historically considered a "trash" tree (it was burned after clearcutting). However, a substance in its bark taxol was recently identified as one of the most promising treatments for ovarian and breast cancer.

Agricultural Value

There are an estimated 80,000 edible plants in the world. Humans depend upon only 20 species of these plants, such as wheat and corn, to provide 90% of the world's food. Wild relatives of these common crops contain essential disease-resistant material. They also provide humans with the means to develop new crops that can grow in inadequate lands such as in poor soils or drought-stricken areas to help solve the world hunger problem.

Ecological Value

Plant and animal species are the foundation of healthy ecosystems. Humans depend on ecosystems such as coastal estuaries, prairie grasslands, and ancient forests to purify their air, clean their water, and supply them with food. When species become endangered, it is an indicator that the health of these vital ecosystems is beginning to unravel.

The U.S. Fish and Wildlife Service estimates that losing one plant species can trigger the loss of up to 30 other insect, plant and higher animal species. The northern spotted owl, listed as threatened in 1990, is an indicator of the declining health of the ancient forests of the Pacific Northwest. These forests are the home to over 100 other old-growth dependent species, which are at risk due to decades of unsustainable forest management practices.

Pollution off the coast of Florida is killing the coral reefs along the Florida Keys, which serve as habitat for hundreds of species of fish. Commercial fish species have begun to decline, causing a threat to the multi-million dollar tourism industry, which depends on the quality of the environment.

Commercial Value

Various wild species are commercially raised, directly contributing to local and regional economies. Fore example, commercial and recreational fishing provides many jobs and contribute to the annual personal income. This industry and way of life, however, is in trouble as salmon decline due to habitat degradation from dams, clearcutting, and overgrazing along streams. Freshwater mussels which are harvested, cut into beads, and used to stimulate pearl construction in oysters form the basis of a thriving industry. Unfortunately, 43% of the freshwater mussel species are currently endangered or extinct.

Recreational Value

Plant and animal species and their ecosystems form the basis of multi-billion dollar, job-intensive tourism industry. They also supply recreational, spiritual, and quality-of-life values as well. Each year millions of people participate in wildlife-related recreation including observing, feeding, and photographing wildlife. People spend billions annually on travel, lodging, equipment, and food to engage in non-consumptive wildlife recreation. For each country, its national heritage of biological diversity is an invaluable and irreplaceable resource. Our quality of life and that of future generations depends on our preservation of plant and animal species.

Recovery Plans

Recovery plans are designed to reverse the decline of a threatened or endangered species and eventually bring the population to a self-sustaining level. Each plan should include:

1. A description of the species' current situation, including any relevant scientific data;
2. A recovery objective (for example, a target population number), and a list of criteria for indicating when the objective has been achieved;
3. An implementation schedule, including priorities of tasks and cost estimates;
4. An appendix identifying appropriate external reviews of the plan, and any additional pertinent information.

A recovery plan may include a myriad of different options including reintroduction, habitat acquisition, captive propagation, habitat restoration and protection, population assessments, research and technical assistance for landowners, and public education. Unfortunately, implementation of a recovery plan is not mandatory, so once it is finalised, the plan may just collect dust on a shelf.

The actual process of preparing the recovery plan varies from circumstance to circumstance. Related organisations can provide some guidance for plan development by setting priorities or arbitrating disputes between individual team members. During and after the drafting process, independent peer review of the plan may be solicited. In addition, there is a public comment period before the plan is finalised and implemented.

Critics of recovery planning often say that the process either doesn't work, costs too much, or both. Endangered Species Acts of various countries have been quite successful in their mission of halting the decline of endangered and threatened species. Recovery planning has clearly played a role in this success. But very few species have actually been recovered and de-listed. Recovery takes a long time because it must reverse a decline that has occurred over the past two centuries.

The recovery period may depend on the status of the species population, the gestation rate of a species, or other biological factors. However, the length of recovery time also depends largely on how quickly an effective recovery plan is developed and implemented. This is largely affected by budget constraints, political pressure, and limited scientific data.

All recovery planning should include an element of 'recovery plan guidance', which details how the plan should be implemented to recover the species. A rational, scientific evaluation of survival and recovery goals is needed. These changes would help make recovery plans quantifiable and based on principles of conservation biology.

The resulting meaningful recovery plans will make the laws for protecting endangered species much more effective and lead to better recovery of species. In addition, if recovery plans identify the types of activities that are likely to violate the laws, predictability will be increased and controversy will diminish when specific projects are evaluated concerning their impact on listed species and their habitat.

Ways to Help Endangered Species

Endangerment is a broad issue, one that involves the habitats and environments where species live and interact with one another. Although some measures are being taken to help specific cases of endangerment, the universal problem cannot be solved until humans protect the natural environments where endangered species dwell.

To survive from extinction endangered species need the help of human beings. Here are some ways that one can get involved:

Conserve Habitats

— One of the most important ways to help threatened plants and animals survive is to protect their habitats permanently in national parks, nature reserves or wilderness areas. There they can live without too much interference from humans. It is also important to protect habitats outside reserves such as on farms and along roadsides.

— You can visit a nearby national park or nature reserve. Some national parks have special guided tours and walks for kids. Talk to the rangers to find out whether there are any threatened species and how they are being protected. You and your friends might be able to help the rangers in their conservation work.

— When you visit a national park, make sure you obey the wildlife code: follow fire regulations; leave your pets at home; leave flowers, birds' eggs, logs and bush rocks where you find them; put your rubbish in a bin or, better still, take it home.

— If you have friends who live on farms, encourage them to keep patches of bush as wildlife habitats and to leave old trees standing, especially those with hollows suitable for nesting animals.

— Some areas have groups which look after local lands and nature reserves. They do this by removing weeds and planting local native species in their place. You could join one of these groups, or even start a new one with your parents and friends. Ask your local parks authority or council for information.

— By removing rubbish and weeds and replanting with natives you will allow the native bush to gradually regenerate. This will also encourage native animals to return.

11

Policies and Programmes for Biodiversity Conservation

Implementation of International Conventions

International agreements have undoubtedly made a significant contribution towards to the preservation of biodiversity across the region. The most important of these is the Convention on Biological Diversity, which was signed by the majority of nations attending the Earth Summit in 1992. These, and others, are described below.

Convention on Biological Diversity (CBD)

The Convention on Biological Diversity (CBD) calls for increased international cooperation in the fight to conserve biodiversity across the globe, and provides a framework for individual countries to develop national strategies and programmes to this effect. The convention encompasses all ecosystems, species and genetic resources, and addresses both traditional conservation efforts and the sustainable exploitation of resources for commercial and economic benefit. Given its comprehensive scope, the CBD has led to action at both the national and international level.

National Action: Recognising that the most effective action will generally take place at the national level, the

CBD requires governments to develop and implement national strategies and programmes to conserve biodiversity, which would include efforts to: monitor biodiversity levels; integrate biodiversity concerns into national decision-making; adopt economic and social measures that encourage conservation and sustainable natural resources use; and support efforts by local populations to adopt more sustainable practices. In response, as of March 1999, some 36 countries of the region had ratified the Convention and a total of 21 countries have now prepared, or are close to preparing, national strategies and action plans.

International Action: At the international level, the CBD's main objective is to ensure that all benefits arising from the use of genetic resources are both fair and equitable. To achieve this, it seeks to promote cooperation to strengthen human resources and institutional capabilities worldwide, particularly in developing countries, through joint programmes for research and technology development, as well as exchange of information and expertise between participating countries. It also seeks to identify and provide funding to developing countries to achieve these aims, primarily by a financing mechanism currently being operated under GEF.

In order to promote international cooperation, the third Conference of the Parties (COP) of the CBD in November 1996, launched a major programme called the "Biotrade Initiative". The initiative consists of three complementary components:

— a country programme under which opportunities and constraints for the development of a sustainable bio-resource industry will be assessed in countries, and 'bio-partnership' will be facilitated.

— market research and policy analysis to include issues such as Intellectual Property Right (IPR), technology transfer, and benefit sharing mechanisms; and

— dissemination and exchange of country programmes' experiences through the establishment of Internet and communication services.

The Biotrade country programme represents the most comprehensive part of the initiative. It analyses opportunities and constraints for the development of a sustainable bio-resource industry. To capture opportunities and solve identified problems, the country programmes develop proactive strategies, focusing on bio-business development, bio-partnerships, sustainable use, conservation, and benefit sharing incentives. The other two components are designed to systematically compile and analyse market data and policy issues. Information thus gathered will be disseminated through a web site, publications and briefings, which it is hoped will form the basis for a more transparent understanding of market dynamics and trends, market barriers, trade and investment flows, property right regimes, and bio-business development, as well as conservation, sustainable use and benefit sharing.

Guidance about the Biotrade Initiative is generally provided by an national advisory panel, which will bring together representatives of the private sector, local and indigenous communities, NGO's, academic institutes, governments and intergovernmental organisations. Another major initiative under CBD is the ongoing negotiations concerning Biosafety protocol. These negotiations reflect the growing concerns about the potential risks posed by living modified organisms produced by modern biotechnology. Many countries with modern biotechnology industries do have domestic legislation, however, there are no binding international agreements addressing situations where living modified organisms cross national borders. The negotiations, therefore, focus on transboundary movements, and include consideration of informed agreement procedures that will enable governments to control imports and refuse those that are unwanted.

Accidental releases of organisms are also being addressed in the protocol. A major concern is that many developing countries lack the technical, financial and institutional capacity to address biosafety. They need greater capacity for assessing and managing risks, establishing adequate information systems, and developing expert human resources in biotechnology.

Initiatives are also being promoted at subregional level in South Asia, Southeast Asia and South Pacific to assist countries in fulfilling the objectives of CBD as well as help conserve biodiversity. For example, with the assistance of European Union, ASEAN has established the ASEAN Regional Centre for Biodiversity Conservation (ARCBC) which serves as the main focal point for networking and institutional linkage among ASEAN Member Countries (AMCs) and between ASEAN and European Union (EU) partner organisations, to enhance the capacity of ASEAN in promoting biodiversity conservation. The EU provides the means for networking, applied research, training and technical assistance, whilst ASEAN provides office space, facilities and support personnel. The Department of Environment and Natural Resources (DENR) in the Philippines is the project's executing agency.

Convention on International Trade in Endangered Species of Wild Fauna and Flora (CITES)

CITES, which regulates the import and export of endangered species of wild flora and fauna, has become a very effective instrument for countering the loss of biodiversity caused by international trade, although there is still work to be done in Asia and the Pacific to improve its implementation and enforcement. The treaty operates through the issuance and control of export and import permits for species that can withstand current rates of exploitation, but prohibits trade in those facing extinction. As of March 1999, 25 countries from Asia and the Pacific had ratified the Convention.

Under this agreement, each member country is responsible for the implementation of CITES within its own jurisdiction, including the appointment of at least one Management Authority, and one Scientific Authority. Several countries have already taken moves to effectively implement the treaty. For example: in India, the Ministry of Environment and Forest established the National Coordination Committee (NCC) in 1995 which has been promoting effective inter-departmental coordination for the control of illegal trade in wildlife and related products; in Hong Kong, China, the Agriculture and Fisheries Department (AFD) launched a public awareness campaign about trade in endangered species; and in Thailand, the Thai Management Authority, in cooperation with the CITES Secretariat, has developed an Orchid Identification Guide to combat the problem of trade in rare orchid specimens.

Convention on Wetlands

The Ramsar Convention is aimed at the protection and conservation of internationally significant wetlands. Signatories are required to adopt the Ramsar Convention Strategic Plan 1997-2002, and to thereafter work with Wetlands International, at both the national and regional level, to implement the Convention. The main challenge for Ramsar signatories is to maintain the ecological character of their listed sites and all their wetlands through conservation and wise use. As of March 1999, 18 countries in the region had ratified the Convention. At the national level, collaboration is achieved through a combination of: development of national wetland strategies and policies; training programmes; wetland surveys; public education and awareness building; and demonstration management of selected wetland sites.

At the regional level, collaboration is achieved through the implementation of multicountry projects and regional wetland inventories. In 1996, an international non-binding framework, the Asia-Pacific Migratory Waterbird Conservation Strategy: 1996-2000, was endorsed by contracting parties to the Ramsar convention to promote the conservation of migratory waterbirds and their wetland habitat in the Asian and Pacific Region. One of the Strategy's priorities is the establishment of networks of sites of internationally important wetland habitats for three groups of waterbird species: shorebirds; cranes; and anatidae.

There is already an East Asian Australasian Shorebird Reserve Network, launched in 1996, under which an Action Plan for the Conservation of Migratory Shorebirds in Asia and the Pacific 1998-2000 guides the work until the end of 2000. A Shorebird Working Group of Experts from Australia, People's Republic of China, Japan, the Philippines and the Russian Federation coordinates the Action Plan. In the South Pacific, a number of tangible achievements in the implementation of the Convention have been noted. Potential Ramsar sites have been documented in Solomon Islands and Vanuatu, and a second Ramsar site has been designated for Papua New Guinea. An inventory of freshwater fish has also been conducted, revealing that far greater species richness is present on certain islands than was thought only 10 years ago.

Convention on the Conservation of Migratory Species of Wild Animals (CMS)

The Convention on the Conservation of Migratory Species of Wild Animals aims to conserve terrestrial, marine and avian migratory species throughout their route or range. It is one of a small number of inter-governmental treaties concerned with the conservation of migratory wildlife and their habitats on a global scale. Seven countries from Asia and the Pacific are parties to the convention, all of whom work together to conserve migratory species and their habitat by providing strict protection for the endangered migratory species; by concluding multilateral agreements for the conservation and management of migratory species; and, by undertaking cooperative research activities.

Amongst the multilateral agreements formed under CMS, the African-Eurasian Migratory Waterbird Agreement (AEWA) is the largest. Worldwide, it covers 172 species of birds that are ecologically dependent on wetlands for at least part of their annual cycle, including many species of pelicans, storks, flamingos, swans, geese, ducks and waders. Other important agreements under CMS that concerns parts of the Asia are the Memorandum of Understanding Concerning Conservation Measures for the Slender-billed Curlew, and the Memorandum of Understanding Concerning Conservation Measures for the Siberian Crane.

Conservation of Ecosystems, Species and Genes

Protected Areas System

At present, about 2.4 million square kilometres of the region's surface area is officially designated as protected, or about five per cent of the total area. In all, there are over 4 000 protected areas in the region, more than two-thirds of which are in Northeast and Southeast Asia; the majority of these are in Indonesia and People's Republic of China. In terms of areal extent, People's Republic of China and the eastern regions of the Russian Federation contain almost half of the officially designated protected area in the region. As a rule of thumb, it has been recommended that country's should aim to designate at least 10 per cent of their territory as protected.

Few countries of the region have been able to reach this benchmark. However, it should be emphasised that percentage share of protected areas is only a very crude statistic of the effectiveness of ecosystem coverage in any one country, and one which does not reflect the relative diversity of the areas protected. No doubt considerable achievements have been made in the region through designation and management of protected areas. However, there are some serious shortcomings with current efforts. For example, protected areas do not always target sites of high biological diversity. Moreover, many protected areas, which are officially decreed as such on paper, are simply not protected in practice, due to reasons such as under-funding and/or understaffing of natural resource agencies.

The lack of commitment by governments to implement protected area legislation is another shortcoming, as a result of which, officially designated reserves are often subjected to agricultural development, mining, extensive poaching, and other forms of degradation. An example is the Narayan Sarovar Sanctuary in India, which is home to a rich assembly of wildlife, including wolves, desert cats, and the largest known population of Indian gazelle. In 1995, this reserve system was turned over to mining companies eager to harvest the coal, bauxite, and limestone deposits found in the area.

Problems also arise when strictly defined borders conflict with the cultural and economic interests of local communities. An example is the current conflict between the Government of the Philippines and inhabitants in and around Mt. Kitanglad National Park, a biodiversity rich area in the southern part of the country. The Government's plans for the area have been strongly opposed by mountain farmers, who object to being forced to switch from traditional monocropping practices into agroforestry, which is perceived as less profitable.

Botanical Gardens and Zoological Parks: Botanical gardens and zoological parks play an important role in both in situ and ex situ conservation of biological diversity. Since botanical gardens and zoological parks store myriads of plant and animal accessions, these represent a vast conservation resource of stored and managed biodiversity. In addition, botanical gardens have also played a crucial role in the development, designation, care and management of conservation areas through activities such as: habitat restoration; wild plant population research, recovery or management; individual species recovery programmes; development and maintenance of databanks; identification and development of economically important species in commercial horticulture, forestry and agriculture and bioprospecting, etc.

Five countries-Australia, People's Republic of China, India, Japan and the Russian Federation-house the majority of botanical gardens in the region. A recent initiative to link botanical gardens within a coordinated global network is the founding of Botanic Gardens Conservation International (BGCI), whereby member-countries are working towards the implementation of a worldwide Botanic Gardens Conservation Strategy and Action Plan for plant conservation, providing technical guidance, data and supporting for botanical gardens, and helping create or strengthen national and regional networks of gardens in many countries, including Australia, People's Republic of China, India and Indonesia.

Gene Banks: The conventional solution to the conservation of plant germplasm has been the establishment of genebanks, which serve as storehouses for germplasm collections, including seed banks, field genebanks, and tissue collections in culture. In Asia and the Pacific, they are commonly integrated with botanical gardens and have been an important source of material for plant breeding programmes and other research activities, thereby forming a basic element in biodiversity conservation programmes in the region.

Maintaining botanical gardens and gene banks, however, is costly, time-consuming and labour-intensive. For example, the current estimated total annual cost of maintaining samples in gene banks is US$50 per accession. Many countries in the region do not have the facilities and are hard-pressed for operating funds; therefore, they cannot maintain samples of a number of species under optimal physical conditions.

Bioregional Management

Policymakers and environmental managers in several countries of the region have begun to adopt an integrated, "total ecosystem" strategy for regulating activities which adversely affect biodiversity. Such an approach, whether on land or at sea, can be used to balance conservation needs with the economic and social demands of the people living within or close to the habitats which are being protected. Integrated coastal zone management programmes provide a good example of the concept of bioregional management.

By far the largest application of bioregional management within the region is the Regional Seas Programme initiated by UNEP, an action-oriented strategy focusing on the mitigation or elimination of both the causes and consequences of environmental degradation. It has a comprehensive, integrated, and results-oriented approach towards combating environmental problems through rational management and development of marine and coastal areas. Each subregion is required to formulate their own action plans to promote the parallel development of intra-regional legal agreements. Another significant example of bioregional management is the International Coral Reef Initiative (ICRI), created to protect coral reefs and the associated ecosystems such as sea grass beds and mangroves.

ICRI currently involves Australia, Japan, and the Philippines and seeks to implement those components of Agenda 21 which call on states to take a special care of marine ecosystems exhibiting high levels of biodiversity and productivity. Establishment of transboundary-protected areas is also taking place in many parts of the region in order to promote bioregional management. For example, the Mekong River Commission has been tasked to coordinate activities related to the use and development of water and associated resources of the Lower Mekong Basin, and the concept of transboundary conservation within the Hindu-Kush Himalayan area is being facilitated by international organisations such as ICIMOD and WWF.

Another good example of a transboundary reserve is the Turtle Islands group, a well-defined rookery of green turtles shared by nine islands, of which six are within Philippine territory. Within this group, the Philippines, Malaysia and Indonesia are all cooperating with one another in conducting scientific and sociological studies to promote turtle conservation. For example, the

Philippine-Sabah Turtle Islands and Berau, Indonesia support the only major green sea turtle breeding populations in the world. These countries also collaborate in tagging and monitoring activities, hatcheries operations, and monitoring of traffic of turtle eggs between islands.

Finally, in the Brunei Darussalam/Indonesia/Malaysia/Philippines East Asian Growth Area (BIMPEAGA) subregion, the Working Group on Environmental Management are also implementing various projects with the involvement of the private sector. Several inter-governmental regional agreements such as the SACEP in South Asia, ASEAN in Southeast Asia and SPREP in the South Pacific also deal with aspects of biodiversity conservation. For example, SACEP's Strategy and Programme includes regional cooperation in wildlife conservation and genetic resources and conservation of corals, mangroves, deltas, and coastal areas.

In addition, ASEAN established a regional framework for the promotion of conservation and sustainable use of heritage areas and endangered species. In the South Pacific biodiversity-related projects are included under the South Pacific Biodiversity Conservation Programme (SPBCP). Asia and the Pacific is extremely rich in biodiversity, and possesses an immense variety of ecological habitats and climatic conditions. Seven out of seventeen mega diversity countries of the world are located in the region, which not only have a wealth of biodiversity, but are also noted for their high rates of species endemism.

The region's biodiversity, however, has been under serious threat as a result of habitat loss, over-exploitation of resources, and the introduction of exotic species. A range of economic and social pressures on the environment exacerbates these problems. According to the IUCN Red List, more than 10 000 higher plant species and over 3

000 vertebrate animal species in Asia and the Pacific are threatened. As natural habitats rapidly shrink, over a thousand birds, about a thousand mammals, and several hundred reptiles and fish species are at risk. Genetic diversity is also diminishing, particularly in relation to domesticated crops and livestock.

In the short-term, this means that existing genetic resources are less able to adapt to conditions of stress, for example, in relation to pathogens or pests, drought or temperature extreme. In the long-term, it means the loss of well-adapted genetic varieties. In an effort to preserve biodiversity in the region, policies and programmes have been promoted at the national level, and coordination of these programmes has been promoted at the regional and international level. Many countries in the region have prepared national strategies and action plans in response to international conventions such as the Convention on Biological Diversity.

These strategies and plans emphasise actions in areas such as: the enhancement of institutional capacity; improved management systems for in situ and ex situ conservation; sustainable utilisation of biodiversity by mainstreaming within other sectoral strategies; enhancing public awareness; research, surveys, and monitoring; partnerships with stakeholders, including increasing community participation in the management of biodiversity; and the promotion of national and international cooperation. All these components, if properly implemented, could have significant bearing on the region's biodiversity.

However, inadequate human and financial resources, particularly in the developing countries of the region, are hampering implementation of these strategies and action plans. Protected area systems in the region are still limited in extent, as they constitute only five per cent of the total area, against an IUCN guideline of 10 per cent. Moreover, many important habitats are either un-represented or under-represented in the present system of protected areas. The coverage of wetlands and marine ecosystems, for example, is still extremely limited. Clearly, there remains considerable scope for expanding the network further, through conservation and restoration of natural areas or traditionally maintained land and/or seascapes.

Another deficiency includes a preponderance of small and fragmented areas, which jeopardizes their integrity and long-term viability. In terms of genetic resources, the concept of intellectual property rights and the lack of a clear, multilateral system for these rights distract stakeholders in countries of the region from the task of conservation of these resources. Governments in Asia and the Pacific should support the right of subsistence farmers to save and adopt the seeds they plant, which is arguably the most important mechanism for sustaining agrobiodiversity.

It is important to build institutional capacity in this regard, and to promote measures such as bioprospecting that could contribute both to conservation of biodiversity and enhancement of resources for conservation. Research, training and information to help expand the capacity for conservation of genes, species and ecosystems also remains vital. Biotechnology and genetic engineering also appear to have significant potential for generating income amongst the region's poor, however, these are new fields, and much remains to be found out about the potential environmental and health risks from the interaction of genetically engineered or modified organisms with natural ecosystems, and strong regulatory and legal frameworks need to be introduced to protect against these impacts.

Finally, there is a need to provide appropriate incentives to individuals, institutions and industries that depend directly on biodiversity for their well-being; to invest in strategies that will help conserve ecosystems whilst promoting and sustaining developing country economies. There is a simultaneous need to involve these various stakeholders in policy and decision-making through multi-stakeholder programmes, environmental steering committees, public consultations and public awareness programmes.

12

The Convention on Biological Diversity

Throughout history, human societies have established rules and customs to keep the use of natural resources within limits in order to avoid long-term damage to the resource. Aspects of biodiversity management have been on the international agenda for many years, although early international environmental treaties were primarily concerned with controlling the excess exploitation of particular species. The origins of modern attempts to manage global biological diversity can be traced to the United Nations Conference on Human Environment held in Stockholm in 1972, which explicitly identified biodiversity conservation as a priority.

The Action Plan in Programme Development and Priorities adopted in 1973 at the first session of the Governing Council of UNEP identified the "conservation of nature, wildlife and genetic resources" as a priority area. The international importance of conservation was confirmed by the adoption, in the same decade, of the Convention on Wetlands, the World Heritage Convention, the Convention on International Trade in Endangered Species, and the Convention on Migratory Species as well as various regional conventions.

By the 1980s, however, it was becoming apparent that traditional conservation alone would not arrest the decline of biological diversity, and new approaches would be needed to address collective failure to manage the human environment and to achieve equitable human development. Important declarations throughout the 1980s, such as the World Conservation Strategy and the resolution of the General Assembly of the United Nations on the World Charter for Nature, stressed the new challenges facing the global community.

In 1983 the General Assembly of the United Nations approved the establishment of a special independent commission to report on environment and development issues, including proposed strategies for sustainable development. The 1987 report of this World Commission on Environment and Development, entitled Our Common Future, argued that "the challenge of finding sustainable development paths ought to provide the impetus—indeed the imperative—for a renewed search for multilateral solutions and a restructured system of cooperation. These challenges cut across the divides of national sovereignty, of limited strategies for economic gain, and of separated disciplines of science."

A growing consensus was emerging among scientists, policy-makers and the public, that the biosphere had to be seen as a single system, and that its conservation required multilateral action, since global environmental problems cannot by definition be addressed in isolation by individual States, or even by regional groupings. By the end of the 1980s, international negotiations were under way that would lead to the United Nations Conference on Environment and Development, held in Rio de Janeiro in June 1992. At this pivotal meeting, Agenda 21, the Rio Declaration on Environment and Development, and the Statement of Forest Principles were adopted, and both the

United Nations Framework Convention on Climate Change and the Convention on Biological Diversity were opened for signature.

During the same period there was an increasing interest in international mechanisms for environmental funding. With the debt crisis, commercial flows for development had become scarce, and the role of multilateral assistance had assumed greater importance in discussions on financial flows and debt rescheduling. Simultaneously, concern with new funding for environmental issues was growing—the Brundtland Report argued for a significant increase in financial support from international sources; the 1987 Montreal Protocol on Substances that Deplete the Ozone Layer established a financial mechanism to provide financial and technical assistance to eligible Parties for the phasing-out of chlorofluorocarbons (CFCs); and the concept of debt-for-nature swaps, that would promote "win-win" situations allowing developing countries to ease their debt burdens and finance environmental protection, was being examined.

A number of proposals for funds and mechanisms were made. Donor country readiness to increase the supply of funds was low and their willingness to support new international agencies even lower, but nevertheless recognition of the principle that additional environment-related funding would have to be provided to developing countries was emerging. During 1989 and 1990 discussions took place within the framework of the World Bank's Development Committee on a new funding mechanism for the environment.

At the end of 1990 agreement was reached on the establishment of the Global Environment Facility under a tripartite agreement between the World Bank, UNDP and UNEP. The GEF would be a pilot initiative for a three-year period to promote international cooperation and to foster action to protect the global environment. The grants and concessional funds disbursed would complement traditional development assistance by covering the additional costs (also known as "agreed incremental costs") incurred when a national, regional or global development project also targets global environmental objectives.

The GEF was given four focal areas, one of which was to be biological diversity. One of the first initiatives taken under the pilot phase was to support preparation of Biodiversity Country Studies in twenty-four developing countries and countries in transition. The primary objective of the Biodiversity Country Studies was to gather and analyse the data required to drive forward the process of developing national strategies, plans, or programmes for the conservation and sustainable use of biological diversity and to integrate these activities with other relevant sectoral or cross-sectoral plans, programmes, or policies. This anticipated the provisions of key articles of the Convention on Biological Diversity, in particular the requirements in Article 6 for each country to have a national biodiversity strategy and to integrate the conservation and sustainable use of biodiversity into all sectors of national planning and in Article 7 to identify components of biological diversity important for its conservation and sustainable use.

The World Conservation Union (IUCN) had been exploring the possibilities for a treaty on the conservation of natural resources, and between 1984 and 1989 had prepared successive drafts of articles for inclusion in a treaty. The IUCN draft articles concentrated on the global action needed to conserve biodiversity at the genetic, species and ecosystem levels, and focused on in situ conservation within and outside protected areas. It also included the provision of a funding mechanism to share the conservation burden between the North and the South.

In 1987 the Governing Council of UNEP established an Ad Hoc Working Group of Experts on Biological Diversity to investigate "the desirability and possible form of an umbrella convention to rationalise current activities in this field, and to address other areas which might fall under such a convention." The Group of Experts concluded that while existing global and regional conventions addressed different aspects of biological diversity, the specific focus and mandates of these conventions did not constitute a regime that could ensure global conservation of biological diversity. On the other hand, it also concluded that the development of an umbrella agreement to absorb or consolidate existing conventions was legally and technically impossible.

By 1990 the Group had reached a consensus on the need for a new global treaty on biological diversity, in the form of a framework treaty building on existing conventions. The scope of such a convention was broadened to include all aspects of biological diversity, including in situ and ex situ conservation of wild and domesticated species, sustainable use of biological resources, access to genetic resources and to relevant technology, including biotechnology, access to benefits derived from such technology, safety of activities related to living modified organisms, and provision of new and additional financial support.

In February 1991 the Group of Experts became the Intergovernmental Negotiating Committee for a Convention on Biological Diversity (INC). The INC held seven negotiating sessions, aiming to have the Convention adopted in time for it to be signed by States at the Earth Summit in June 1992. The relationship between the objectives of the Convention and issues relating to trade, to agriculture and to the emerging biotechnology sector were key issues in the minds of the negotiators. Part of the novelty of the Convention on Biological Diversity lies in the recognition that, to meet its objectives, the Convention would need to make sure that these objectives were acknowledged and taken account of by other key legal regimes.

These included the trade regime that would enter into force in 1994 under the World Trade Organisation; the FAO Global System on Plant Genetic Resources, in particular the International Undertaking on Plant Genetic Resources adopted in 1983; and the United Nations Convention on the Law of the Sea which was concluded in 1982 and would enter into force in 1994. Those involved in negotiating the Convention on Biological Diversity, as well as those involved in the parallel negotiations on the United Nations Framework Convention on Climate Change, were consciously developing a new generation of environmental conventions.

These conventions recognised that the problems they sought to remedy arose from the collective impacts of the activities of many major economic sectors and from trends in global production and consumption. They also recognised that, to be effective, they would need to make sure that the biodiversity and climate change objectives were taken into account in national policies and planning in all sectors, national legislation and relevant international legal regimes, the operations of relevant economic sectors, and by citizens of all countries through enhanced understanding and behavioural changes. The text of the Convention was adopted in Nairobi on 22 May 1992, and between 5 and 14 June 1992 the Convention was signed in Rio de Janeiro by the unprecedented number of 156 States and one regional economic integration organisation.

The early entry into force of the Convention only 18 months later, on 29 December 1993, was equally unprecedented, and by August 2001 the Convention had 181 Contracting Parties.

Objectives and Approach of the Convention

The objectives of the Convention on Biological Diversity are "the conservation of biological diversity, the sustainable use of its components, and the fair and equitable sharing of the benefits arising out of the utilisation of genetic resources" (Article 1). These are translated into binding commitments in its normative provisions, contained in Articles 6 to 20. A central purpose of the Convention on Biological Diversity, as with Agenda 21 and the Convention on Climate Change, is to promote sustainable development, and the underlying principles of the Convention are consistent with those of the other "Rio Agreements".

The Convention stresses that the conservation of biological diversity is a common concern of humankind, but recognises that nations have sovereign rights over their own biological resources, and will need to address the overriding priorities of economic and social development and the eradication of poverty. The Convention recognises that the causes of the loss of biodiversity are diffuse in nature, and mostly arise as a secondary consequence of activities in economic sectors such as agriculture, forestry, fisheries, water supply, transportation, urban development, or energy, particularly activities that focus on deriving short-term benefits rather than long-term sustainability.

Dealing with economic and institutional factors is therefore key to achieving the objectives of the Convention. Management objectives for biodiversity must incorporate the needs and concerns of the many stakeholders involved, from local communities upward. A major innovation of the Convention is its recognition that all types of knowledge systems are relevant to its objectives. For the first time in an international legal instrument, the Convention recognises the importance of traditional knowledge—the wealth of knowledge, innovations and practices of indigenous and local communities that are relevant for the conservation and sustainable use of biological diversity.

It calls for the wider application of such knowledge, with the approval and involvement of the holders, and establishes a framework to ensure that the holders share in any benefits that arise from the use of such traditional knowledge. The Convention therefore places less emphasis on a traditional regulatory approach. Its provisions are expressed as overall goals and policies, with specific action for implementation to be developed in accordance with the circumstances and capabilities of each Party, rather than as hard and precise obligations. The Convention does not set any concrete targets, there are no lists, no annexes relating to sites or protected species, thus the responsibility of determining how most of its provisions are to be implemented at the national level falls to the individual Parties themselves.

Institutional Elements of the Convention

The Convention establishes the standard institutional elements of a modern environmental treaty: a governing body, the Conference of the Parties; a Secretariat; a scientific advisory body; a clearing-house mechanism and a financial mechanism. Collectively, these translate the general commitments of the Convention into binding norms or guidelines, and assist Parties with implementation. Because the Convention is more than a framework treaty, many of its provisions require further collective elaboration in order to provide a clear set of norms to guide States and stakeholders in their management of biodiversity.

Development of this normative basis centres around decisions of the Conference of the Parties (COP), as the governing body of the Convention process. The principal function of the COP is to regularly review implementation of the Convention and to steer its development, including establishing such subsidiary bodies as may be required. The COP meets on a regular basis and held five meetings in the period 1994 to 2000. At its fifth meeting (2000) the COP decided that it would henceforth meet every two years. The Subsidiary Body on Scientific, Technical and Technological Advice (SBSTTA) is the principal subsidiary body of the COP.

Its mandate is to provide assessments of the status of biological diversity, assessments of the types of measures taken in accordance with the provisions of the Convention, and advice on any questions that the COP may put to it. SBSTTA met five times in the period 1995 to 2000 and, in the future, will meet twice in each two-year period between meetings of the COP. The principal functions of the Secretariat are to prepare for and service meetings of the COP and other subsidiary bodies of the Convention, and to coordinate with other relevant international bodies. The Secretariat is provided by UNEP and is located in Montreal, Canada.

The Convention provides for the establishment of a clearing-house mechanism to promote and facilitate technical and scientific cooperation (Article 18). A pilot phase of the clearing-house mechanism took place from 1996 to 1998 and, following evaluation of this, the COP has approved a clearing-house mechanism strategic plan and a programme of work until 2004. The Convention establishes a financial mechanism for the provision of resources to developing countries for the purposes of the Convention.

The financial mechanism is operated by the Global Environment Facility (GEF) and functions under the authority and guidance of, and is accountable to, the COP. GEF activities are implemented by the United Nations Development Programme (UNDP), UNEP and the World Bank. Under the provisions of the Convention, developed country Parties undertake to provide "new and additional financial resources to enable developing country Parties to meet the agreed full incremental cost of implementing the obligations of the Convention" (Article 20) and, in addition to the provision of resources through the GEF, these Parties may also provide financial resources through bilateral and multilateral channels. The COP is able, if it deems it necessary, to establish intersessional bodies and meetings to carry out work and provide advice between ordinary meetings of the COP.

Cartagena Protocol

The Convention requires the Parties to "consider the need for and modalities of a protocol setting out appropriate procedures, including, in particular, advance informed agreement, in the field of the safe transfer, handling and use of any living modified organism resulting from biotechnology that may have adverse effect on the conservation and sustainable use of biological diversity" (Article 19(3)).

At its second meeting, the COP established a negotiating process and an Ad Hoc Working Group on Biosafety that met six times between 1996 and 1999 to develop a draft protocol. The draft submitted by the Working Group was considered by an Extraordinary Meeting of the COP held in Cartagena, Colombia in

February 1999 and in Montreal, Canada in January 2000, and on 29 January 2000 the text of the Cartagena Protocol on Biosafety to the Convention on Biological Diversity was adopted.

The Protocol was opened for signature during the fifth meeting of the COP in May 2000 where it was signed by 68 States. The number of signatures had risen to 103 by 1 August 2001, and five States had ratified the Protocol. It will enter into force after the fiftieth ratification. The COP will serve as the meeting of the Parties to the Protocol. The meetings will however be distinct, and only Parties to the Convention who are also Parties to the Protocol may take decisions under the Protocol.

Pending the entry into force of the Protocol, an Intergovernmental Committee for the Cartagena Protocol (ICCP) has been established to undertake the preparations necessary for the first meeting of the Parties. The first meeting of the Intergovernmental Committee was held in Montpellier, France in December 2000 and the second in Nairobi, Kenya in September-October 2001.

Decision-Making Process

The activities of the COP have been organised through programmes of work that identify the priorities for future periods. The first medium-term programme of work saw a focus on developing the procedures and modus operandi of the institutions, determining priorities, supporting national biodiversity strategies, and developing guidance to the financial mechanism. At its fourth meeting, the COP adopted a programme of work for its fifth, sixth and seventh meetings, and, at its fifth meeting, approved a longer-term programme of work for SBSTTA, and began the development of a strategic plan for the Convention. The following are the key steps in the decision-making process.

Identification: The programme of work establishes a timetable indicating when the COP will consider in detail biological themes or ecosystems, or specific provisions of the Convention contained in the operative Articles. In addition to such ecosystem based programmes, the COP has addressed a number of key substantive issues in a broadly comprehensive manner. Such issues are collectively known as "cross-cutting issues", and these have an important role to play in bringing cohesion to the work of the Convention by linking the thematic programmes.

Submissions and Compilation of Information: The procedures by which the COP comes to adopt its decisions are broadly similar in each case. Firstly, current activities are reviewed to identify synergies and gaps within the existing institutional framework, or an overview of the state of knowledge on the issue under examination is developed. At the same time, Parties, international organisations, specialist scientific and non-governmental organisations are invited to provide information, such as reports or case studies. This review mechanism is coordinated by the Secretariat, supported in some cases by informal inter-agency task forces or liaison groups of experts.

Preparation of synthesis: The Secretariat then prepares a preliminary synthesis of these submissions for consideration by SBSTTA. Where appropriate the Secretariat may use a liaison group to assist with this. In other cases SBSTTA may have established an ad hoc technical expert group, with members drawn from rosters of experts nominated by Parties, to assist with the preparation of the synthesis. Where appropriate, the Secretariat may also identify relevant networks of experts and institutions, and coordinate their input to the preparation of the synthesis.

Scientific, Technical or Technological Advice: On the basis of the work of the Secretariat, of any ad hoc technical expert group, and the findings of specialist meetings such as the Global Biodiversity Forum, SBSTTA will assess the status and trends of the biodiversity of the ecosystem in question or the relationship of the cross-cutting issue to the implementation of the Convention and develop its recommendation to the COP accordingly.

Supplementary Preparations for the COP: The advice of SBSTTA may be complemented by the work of the Secretariat in the intersessional period between the meeting of the SBSTTA and that of the COP. Such work may comprise issues not within the mandate of the SBSTTA, such as financial and legal matters, development of guidance to the financial mechanism, or relations with other institutions and processes that could contribute to implementation of the future decision of the COP.

Programmes: The COP considers the recommendations of the SBSTTA and any other advice put before it. It will then advise Parties on the steps they should take to address the issue, in light of their obligations under the Convention. It may also establish a process or programme to develop the issue further. Such a programme would establish goals and identify the expected outcomes, including a timetable for these and the means to achieve them. The types of output to be developed could include: guidelines, codes of conduct, manuals of best practice, guidance for the institutions of the Convention, criteria, and so forth. The programme would proceed to develop these products, under the guidance of SBSTTA, and report results to the COP for review.

Obligations on Parties

The Convention constitutes a framework for action that

will take place mainly at the national level. It places few precise binding obligations upon Parties, but rather provides goals and guidelines, and these are further elaborated by decisions of the COP. Most of the commitments of Parties under the Convention are qualified, and their implementation will depend upon the particular national circumstances and priorities of individual Parties, and the resources available to them. Nevertheless, Parties are obliged to address the issues covered by the Convention, the chief of which are outlined in the following sections.

Article 6: National Strategies and Plans

The implementation of the Convention requires the mobilisation of both information and resources at the national level. As a first step, the Convention requires Parties to develop national strategies, plans or programmes for the conservation and sustainable use of biodiversity, or to adapt existing plans or programmes for this purpose (Article 6(a)). This may require a new planning process, or a review of existing environmental management or other national plans.

The Convention also requires Parties to integrate conservation and sustainable use of biodiversity into relevant sectoral or cross-sectoral plans, programmes and policies, as well as into national decision-making (Article 6(b)). This is clearly a more complex undertaking, requiring an assessment of the impacts of other sectors on biodiversity management. It will also require coordination among government departments or agencies. A national biodiversity planning process can identify the impacts and opportunities for integration.

Given the importance of stakeholder involvement in the implementation of the Convention, national planning processes should provide plenty of scope for public consultation and participation. The COP has recommended the guidance for the development of national strategies found in: Guidelines for Preparation of Biodiversity Country Studies (UNEP) and National Biodiversity Planning: Guidelines Based on Early Country Experiences (World Resources Institute, UNEP and IUCN). The financial mechanism has supported 125 countries in the preparation of their national biodiversity strategies and action plans.

Article 7: Identification and Monitoring of Biodiversity

In contrast to some previous international or regional agreements on conservation, the Convention does not contain an internationally agreed list of species or habitats subject to special measures of protection. This is in line with the country-focused approach of the Convention. Instead, the Convention requires Parties to identify for themselves components of biodiversity important for conservation and sustainable use (Article 7).

Information provides the key for the implementation of the Convention, and Parties will require a minimum set of information in order to be able to identify national priorities. Whilst it contains no lists, the Convention does indicate, in Annex I, the types of species and ecosystems that Parties might consider for particular attention. Parties are also required to monitor important components of biodiversity, and to identify processes or activities likely to have adverse effects on biodiversity. The development of indicators may assist Parties in monitoring the status of biological diversity and the effects of measures taken for its conservation and sustainable use.

Article 8: Conservation of Biodiversity in Situ

The Convention addresses both in situ and ex situ conservation, but the emphasis is on in situ measures, i.e.

within ecosystems and natural habitats or, in the case of domesticated or cultivated species, in the surroundings where they have developed their distinctive properties. Article 8 sets out a comprehensive framework for in situ conservation and a Party's national biodiversity planning process should include consideration of the extent to which it currently addresses the following issues.

Protected areas Parties should establish a system of protected areas or areas where special measures are required to conserve biological diversity, covering both marine and terrestrial areas. They are expected to develop guidelines for the selection, establishment and management of these areas, and to enhance the protection of such areas by the environmentally sound and sustainable development of adjacent areas. Regulation and management of biological resources Parties should regulate or manage important components of biological diversity whether found within protected areas or outside them.

Legislation or other regulatory measures should therefore be introduced or maintained to promote the protection of ecosystems, natural and semi-natural habitats and the maintenance of viable populations of species in natural surroundings. Regulation and management of activities Under Article 7 Parties should attempt to identify activities that may be detrimental to biological diversity. Where such activities have been identified, Parties should take steps to manage them so as to reduce their impacts. Rehabilitation and restoration Parties should develop plans and management strategies for the rehabilitation and restoration of degraded ecosystems and the recovery of threatened species.

Alien species Parties should prevent the introduction of, and control or eradicate alien species which threaten ecosystems, habitats, or native species. Living modified organisms Parties should establish or maintain means to manage the risks associated with the use and release of living modified organisms (LMOs) resulting from biotechnology. Parties are thus required to take action at the national level to ensure that LMOs do not cause adverse effects to biodiversity.

Traditional knowledge and practices The Convention recognises that indigenous and local communities embodying traditional lifestyles have a crucial role to play in the conservation and sustainable use of biodiversity. It calls on Parties to respect, preserve and maintain the knowledge, innovations and practices of indigenous and local communities and to encourage their customary uses of biological resources compatible with the conservation and sustainable use of these resources. By this, the Convention acknowledges the significance of traditional knowledge and practices, which should be taken into account in the implementation of all aspects of the Convention.

Article 9: Conservation of Biodiversity Ex Situ

While prioritising in situ conservation, the Convention recognises the contribution that ex situ measures and facilities, such as gene banks, botanic gardens and zoos, can make to the conservation and sustainable use of biological diversity. It specifies that, where possible, facilities for ex situ conservation should be established and maintained in the country of origin of the genetic resources concerned. The Convention does not, however, apply its provisions on access and benefit-sharing to ex situ resources collected prior to the entry into force of the Convention. This is of particular concern to developing countries, from which natural resources have already been removed and stored in ex situ collections, without a mechanism to ensure the sharing of benefits. The issue of the status of ex situ resources is currently being reviewed within the context of the work of the Food and Agriculture Organisation of the United Nations.

Article 10: Sustainable Use

Although the term conservation has sometimes been taken to incorporate sustainable use of resources, in the Convention the two terms appear side by side, and a specific Article of the Convention is devoted to sustainable use. This reflects the view of many countries during the negotiation of the Convention that the importance of sustainable use of resources be accorded explicit recognition. Sustainable use is defined in the Convention as: "the use of components of biological diversity in a way and at a rate that does not lead to the long-term decline of biological diversity, thereby maintaining its potential to meet the needs and aspirations of present and future generations."

The practical implications of this definition in terms of management are difficult to assess. Article 10 does not suggest quantitative methods for establishing the sustainability of use, but sets out five general areas of activity: the need to integrate conservation and sustainable use into national decision-making; to avoid or minimise adverse impacts on biological diversity; to protect and encourage customary uses of biodiversity in accordance with traditional cultural practices; to support local populations to develop and implement remedial action in degraded areas; and to encourage cooperation between its governmental authorities and its private sector in developing methods for sustainable use of biological resources.

Articles 11-14: Measures to Promote Conservation and Sustainable Use

The Convention makes explicit reference to a number of

additional policy and procedural measures to promote conservation and sustainable use. For example, it requires Parties to adopt economically and socially sound incentives for this purpose (Article 11). It also recognises the importance of public education and awareness to the effective implementation of the Convention (Article 13). Parties are therefore required to promote understanding of the importance of biodiversity conservation, and of the measures needed.

Research and training are critical to the implementation of almost every substantive obligation. Some deficit in human capacity exists in all countries, particularly so in developing countries. The Convention requires Parties to establish relevant scientific and technical training programmes, to promote research contributing to conservation and sustainable use, and to cooperate in using research results to develop and apply methods to achieve these goals (Article 12). Special attention must be given to supporting the research and training needs of developing countries, and this is explicitly linked to the provisions on access to and transfer of technology, technical and scientific cooperation and financial resources.

Parties are required to introduce appropriate environment impact assessment (EIA) procedures for projects likely to have significant adverse effects on biodiversity (Article 14). Legislation on EIA will generally incorporate a number of elements, including a threshold for determining when an EIA will be required, procedural requirements for carrying it out, and the requirement that the assessment be taken into account when determining whether the project should proceed. In addition, Parties are required to consult with other States on activities under their jurisdiction and control that may adversely affect the biodiversity of other States, or areas beyond national jurisdiction.

Articles 15-21: Benefits

The Convention provides for scientific and technical cooperation to support the conservation and sustainable use of biological diversity, and a clearing-house mechanism is being developed to promote and facilitate this cooperation. The provisions on scientific and technical cooperation provide a basis for capacity-building activities. For example, the COP has requested the financial mechanism to support a Global Taxonomy Initiative designed, among other things, to develop national, regional and sub-regional training programmes, and to strengthen reference collections in countries of origin.

In addition to general provisions on cooperation, research and training, the Convention includes articles promoting access to the potential benefits resulting from the use of genetic resources, access to and transfer of relevant technology, and access to increased financial resources. The potential benefits for developing country Parties under the Convention arise from the new position on conservation negotiated between developed and developing countries. The extent to which these benefits materialise is likely to be crucial to determining the long-term success of the Convention. Global biodiversity increases toward the tropics, and the Convention gives developing countries, in this zone and elsewhere, an opportunity to derive financial and technical benefits from their biological resources, while the world overall benefits from the goods and services that the biodiversity thus conserved will continue to provide.

Access to genetic resources and benefit-sharing

Before the negotiation of the Convention, genetic resources were considered to be freely available, despite their potential monetary value. However, the approach taken in the Convention is radically different. Article 15 reaffirms the sovereignty of Parties over their genetic resources, and recognises the authority of States to determine access to those resources. While the Convention addresses sovereignty over resources, it does not address their ownership, which remains to be determined at national level in accordance with national legislation or practice. Although the sovereign rights of States over their genetic resources is emphasised, access to genetic resources for environmentally sound uses by scientific and commercial institutions under the jurisdiction of other Parties is to be facilitated.

Since genetic resources are no longer regarded as freely available, the Convention paves the way for new types of regimes governing the relationship between providers and users of genetic resources. It is generally agreed that benefit-sharing should extend not only to the government of the country of origin but also to indigenous and local communities directly responsible for the conservation and sustainable use of the genetic resources in question. National legislation might require bio-prospectors to agree terms with such communities for the use of resources, and this may be all the more crucial where bio-prospectors are seeking to draw upon not only the resources themselves, but also upon the knowledge of these communities about those resources and their potential use.

Access to and transfer of technologies

Under Article 16 of the Convention, Parties agree to share technologies relevant to the conservation of biological diversity and the sustainable use of its components, and also technologies that make use of genetic resources. Technology transfer under the Convention therefore incorporates both "traditional" technologies and biotechnology. Biotechnology is defined in the Convention as: any technological application that uses biological systems, living organisms, or derivatives thereof, to make or modify products or processes for specific use. Technologies which make use of genetic resources are subject to special provisions aimed at allowing the country of origin of the resources to share in the benefits arising out of the development of these technologies.

The Convention makes it a specific requirement that all Parties create a legislative, administrative or policy framework with the aim that such technologies are transferred, on mutually agreed terms, to those providing the genetic resources. This obligation extends to technology protected by patents and other intellectual property rights. More generally, developing country Parties are to have access to technology under terms which are fair and most favourable, including on concessional and preferential terms, where mutually agreed.

Article 16 provides that where relevant technology is subject to an intellectual property right such as a patent, the transfer must be on terms which recognise and are consistent with the adequate and effective protection of the property right. However, it also goes on to provide that Parties are to cooperate in ensuring that intellectual property rights are supportive of, and do not run counter to, the objectives of the Convention.

Financial resources

All Parties undertake to provide financial support and incentives for implementation of the Convention at the national level, in accordance with their capabilities. In addition, developed country Parties agree to make available to developing country Parties, new and additional financial resources to meet "the agreed full incremental costs" of implementing measures to fulfil their obligations. In addition to the financial mechanism mentioned earlier, developed country Parties may provide resources to improve implementation of the Convention through overseas development agencies and other bilateral channels. The Convention explicitly recognises that the extent to which developing country Parties will be able to implement their obligations under the Convention will depend on the developed country Parties fulfilling their obligations to provide resources.

The Convention also acknowledges that economic and social development remains the overriding priority of developing countries, and in this regard recognises the special circumstances and needs of the small island developing States. As a result of both these considerations, developed country Parties are expected to give due consideration to the dependence on, distribution and location of biological diversity within developing countries, in particular small island States and those that are most environmentally vulnerable, such as those with arid and semi-arid zones, coastal and mountainous areas.

Implementation of the Convention

The Convention provides for Parties to present reports to the COP on measures taken to implement the provisions of the Convention and their effectiveness in meeting the objectives of the Convention (Article 26). At its second meeting, the COP decided that the first national reports should focus on implementation of Article 6 of the Convention. This article concerns the need to develop a national biodiversity strategy and action plan, and to ensure that the conservation and sustainable use of biological diversity is integrated with the policies and programmes of other sectors.

The information in these reports was considered by the fourth meeting of the COP, which asked SBSTTA to give advice on the nature of the information required from Parties in order to assess the state of implementation of the Convention. At its fifth meeting, the COP adopted a methodology for national reporting that will enable Parties to provide information on the implementation of all their obligations, as derived from the articles of the Convention and from decisions of the COP that call for action by Parties.

The reporting guidelines will permit Parties to consider the effectiveness of the measures taken and to identify national priorities, national capacity for implementation and constraints encountered. The COP will be able to identify issues that require further scientific or technical investigation, and to identify successes and constraints faced by Parties. In the latter case it will be better placed to decide what steps are necessary to support Parties, and to give appropriate guidance to the financial mechanism, institutions able to assist with capacity development, the Secretariat and to the Parties themselves.

Given the enormous breadth of the issues that the Convention seeks to address, there is need not only for cooperation between Parties, but also to develop institutional links and cooperative relationships with other international bodies. Mechanisms for coordinating these relationships are fundamental to the implementation of the Convention. Each meeting of the COP has reaffirmed the importance it attaches to cooperation and coordination between the Convention and other relevant conventions, institutions and processes, and has invited these to take an active role in the implementation of aspects of the Convention. Equally importantly, the COP has reaffirmed the importance of the role to be played by groups other than States and international bodies.

Non-state actors—national and international non-governmental organisations, scientific bodies, industrial and agricultural associations, and indigenous peoples' organisations, amongst others—have all been called upon to cooperate in scientific assessments, policy development, and implementation of the Convention's work programmes. In particular, as traditional knowledge about conserving and sustainably using biodiversity is central to the development and implementation of the work programmes, cooperation with the holders of traditional knowledge has been particularly emphasised. The institutional structure of the Convention thus extends beyond those institutions established by the process itself.

13

Ecological Networks and Biodiversity Conservation

World's biodiversity occurs largely in ecosystems that have a long association with man. Over the past millennia human activity has increased the number and variety of habitats and biodiversity has increased correspondingly. However over the last century, rapid increases in the use of natural resources and the spread of human habitation have reversed this trend and species and habitats have been put under extreme pressure. As a result habitats are becoming destroyed or fragmented, losing their ability to provide the required goods and services and species are reducing in number and geographic range. Thus to halt this loss of biodiversity, human resource use must become sustainable and integrated with biodiversity conservation.

Habitat Destruction

Habitat destruction is a collective term for a variety of environmental troubles, each of which may have different effects on food web structure and, given that they often act concomitantly, may also interact with each other in unpredictable ways. While a frequent outcome of habitat destruction is species loss, whether the different types of habitat destruction lead to different patterns of species loss remains unknown.

A species is the basic unit of biodiversity. Various species are linked to other species in a variety of critical ways, for example as predators or prey, or as pollinators or seed dispersers. Consequently, the extinction of one species can lead to secondary extinctions in complex ecological networks. The presence of links between species can also lead to community closure after the loss of a species, with the result that this species cannot then be reintroduced.

Food webs have long been used in ecological research and provide a description of the trophic links in a community. In addition to their obvious role as descriptions of community structure, food webs are being used increasingly as the basis for an experimental approach. For example, a food web from Belize was recently use to predict the likely role of indirect effects in structuring the community, these predictions were then tested by a field manipulation.

While food webs may be the most commonly described ecological network, other types of interaction webs are increasingly being investigated that include a variety of trophic and non-trophic interactions such as pollination, seed dispersal, interference competition, habitat or shelter provisioning, and recruitment facilitation or inhibition.

Food webs and other ecological networks have not been widely applied to the field of conservation biology, but given the practical advances being made in food web construction, the theoretical advances, and the ongoing threat of biodiversity loss, now is a good time to begin to use ecological networks as a conservation tool. Conservation biology aims to protect intact native ecosystems and restore degraded ones. Even when these conservation goals focus on protecting one individual threatened or endangered species, there is an increasing recognition that multispecies approaches are essential for success.

Multispecies considerations are also critical for any ecological monitoring or assessment designed to protect key ecosystem services, such as pollution control, pest control, water filtration, and water clarity. At the root of these calls for multispecies approaches to conservation and restoration is an appreciation that species interactions are important. Ironically, conservation, restoration, or monitoring projects rarely collect data on the structure of these interactions or how they change over time and across gradients of human impacts.

Categories of Habitat Destruction

Habitat destruction can be put into four main categories, none of which is mutually exclusive of the others:

Habitat removal leading to and habitat fragmentation

This is probably the most widely known form of habitat destruction and leads to both habitat loss and habitat isolation. The hostility of the matrix between the remaining habitat fragments is likely to affect different species in different ways. For example, the rodents responsible for the spread of Lyme's disease are not affected by fragmentation whereas their competitors and predators are. This difference in matrix permeability leads to both an increase in the incidence of the rodents and in the prevalence of Lyme's disease in humans. This example demonstrates how food web structure can mediate unpredictable effects-here woodland fragmentation indirectly caused an increase in the prevalence of a disease affecting man.

The effect of habitat fragmentation on food web structure is beginning to be addressed, for example, pioneering work by Gilbert et al. revealed that predators are both most affected by fragmentation and that they are most rescued by wildlife corridors. The experiment by Gilbert et al. combined a real community with rigorous experimental design and teased apart the effects of habitat fragmentation on the different trophic levels. However, the work did not explicitly identify links between species; indeed no published work to date has replicated food webs on fragmented habitats.

Hunting

Hunting occurs at all trophic levels; for example canopy trees are removed from tropical rain forests in selective logging programs and vertebrate herbivores and predators are hunted as trophies and for pest control. Insects are rarely hunted, although over-collecting may contribute towards the demise of some butterfly species.

Untold millions of fish, sea turtles, sharks, and manatee have been hunted from the Caribbean Ocean and today no coastal sea can be considered pristine. In addition to potentially driving the hunted species extinct, hunting will obviously also lead to the extinction of species dependent upon the hunted species, for example, host-specific species of fleas and lice.

Introduction of alien species

Aliens pose a significant threat to global biodiversity, second only to habitat loss. They can occur at all trophic levels. Alien plants are a particularly serious threat given their ability to displace native plants, to change the composition of native plant communities, and to alter a range of ecosystem processes such as nutrient cycling and disturbance regimes.

Alien herbivores constitute some of worst agricultural pests, damaging native ecosystems and costing many millions of pounds to control. Alien parasitoids, principally biocontrol agents, have been widely reported from natural habitats around the world and are the subject of considerable current concern. Mammals are one of the most important groups of alien taxa, especially on islands. Alien insect predators can also be very common, for example in Hawaii, 38% of spiders, 23% of predatory beetles, and 14% of Neuroptera are alien species.

While extensive data exist on the distributions of alien species and their impacts on native species such as competitors, prey species, predators, pollinators, and parasites, and even their impact upon ecosystem properties, there is extraordinarily little data on how aliens are accommodated in food webs.

Collapse of Ecosystem

Currently, it is not known whether food webs collapse simply in the opposite order to the way in which they were assembled or whether there is a "first in last out" (FILO) order of collapse. Essentially, these questions ask whether the restoration and recovery of ecosystem function are the mirror image of the processes that lead to decline and collapse. In essence, is there any hysteresis to ecosystem function? Here hysteresis represents the history dependence of the system; thus, if you push on something and it yields, does it spring back completely when released? If it does not, it is exhibiting hysteresis. Or are there situations in which destructive activity focuses on different trophic levels, while recovery always proceeds from the bottom up?

Destructive activity varies with trophic position; thus, plants at the base of a food chain will be lost at rates slightly slower than the average; the decomposers, such as worms, soil mites and bacteria will be lost at the slowest rates. In contrast, the species that feed directly on plants will be lost at a faster rate, while the charismatic tigers and eagles that feed at the top of the food chain are lost at the fastest rate (figure 1).

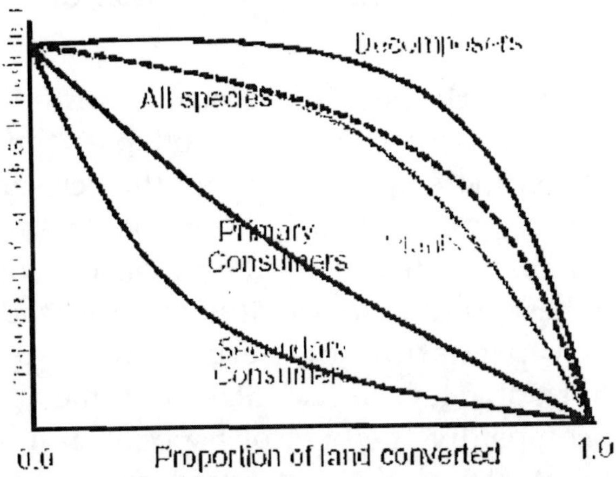

Figure 1. Loss of biological diversity and ecosystem services as natural habitat is eroded

An understanding of how food webs both collapse and reassemble is crucial both in expanding our understanding of how food webs are structured, and also in determining how the ecosystems services they provide will decline as ecosystem functions break down due to species loss.

Examples of ecosystem collapse and reassembly should provide insights into how species loss or addition leads to change in food-web structure. An important way to phrase these questions should be from the perspective of ecosystem and food web hysteresis; specifically, in which ways do the processes of food web/ecosystem assembly resemble those of food web/ecosystem collapse?

The classic example of ecosystem collapse comes from John Terborgh's work on Lago Guri in Venezuela. This lake was flooded in the late 1980s to create a huge artificial lake that would drive a huge hydroelectric dam. Flooding a forested river created a series of islands of different sizes that could be studied as their faunas and floras collapsed. Island biogeography theory predicts that larger islands should conserve a greater diversity of species, which in turn should maintain a higher proportion of ecosystem services.

Terborgh's work uses island size as an adjunct for time in the process of ecosystem collapse; the smaller islands have lost more species and are thus closer to total faunal and .oral "meltdown." The work illustrates that food web collapse is driven by the initial loss of species from the highest trophic levels; this in turn leads to the release from predation of herbivorous species at intermediate trophic levels. The increase in the abundance of the most competitive herbivorous species in turn selects for plants that can withstand herbivory. In the most degraded habitats, leaf cutter ants are only the herbivores that can exploit these plants, which eventually die from defoliation. The general pattern observed here is first a shortening of the web followed by a sequential collapse.

An important contrast to Lago Guri are the studies of Krakatau in the straits between Java and Sumatra. The volcanic eruption in 1883 completely destroyed all life on Krakatau. However, biologists have visited and surveyed the island, and the surrounding newly emergent islands, with an almost decadal frequency; this has provided a unique set of data on how island communities reassemble themselves. While most studies have focused on these data as ways of testing the laws of island biogeography, they represent a rich source of information that could be exploited by the food web community.

When the species living on Krakatau are divided among different trophic levels, we see first a rapid increase in plant species at the base of the web. Herbivorous insects are blown in on the wind, followed by marine and terrestrial bird and bat species. The marine birds feed predominantly on fish, but significantly enhance the soluble nutrients on the island in the guano dumps around their nesting sites. The insectivorous birds and bats feed on the herbivorous insects. Later in the sequence of arrivals are frugivorous birds and bats, which manage to exploit the fruit produced by regurgitated and defecated seeds from the migratory species passing through the islands.

When the buildup of trophic diversity on Krakatau is examined, there are some hystereses like similarities to the patterns observed in Lago Guri. There is an initial rapid assemblage of weakly interacting plant and animal species followed by a slower and more steady increase in diversity at all trophic levels as interactions between species multiply, and intra- and interspecific regulation begin to replace colonisation and persistence as the forces structuring the community.

The Krakatau data can be used to compare potential indices of ecosystem resilience and stability. The most widely used index is mean trophic position; this index is calculated by categorising each species in a trophic level and then calculating the mean for all the species in the community or food web. The index is most widely used in studies of overexploitation of fisheries, where the mean trophic level of species removed from the fishery is estimated.

Almost by definition, this will exclude many species from lower trophic levels from the calculation, as few fisheries actively exploit plankton. A logical extension of this calculation is to calculate the variance, or the standard deviation of trophic position. The standard deviation provides a more sensitive index of food web change than mean trophic level. In particular, it is intriguing that mean trophic level quickly equilibrates as the Krakatau community assembles, while the standard deviation takes longer to stabilise and then does so once the different trophic levels seem to grow at roughly equal rates.

Similar patterns are observed when this approach is applied to the Lago Guri data for community meltdown. In both cases, community change is most rapid when the standard deviation of mean trophic position falls below unity. It would be an interesting exercise to calculate the standard deviation of mean trophic level for other food webs, particularly those that are collapsing. It is probable that food webs with standard deviations of mean trophic level that fall below unity are in the process of collapse.

Species Interactions and Robustness of Ecosystems

With most ecosystems facing mild to extreme stress due to a variety of anthropogenic factors, the question of whether species interactions amplify or ameliorate such perturbations takes on practical importance. Increasingly, conservation biologists and ecologists are called on to help set priorities for conservation and restoration programs.

Understanding and predicting ecosystem robustness to perturbations should be a fundamental part of making sustainable conservation decisions. Food webs and other types of ecological networks can provide a very useful framework for beginning to evaluate the robustness of ecosystems to species loss, and the tendency of complex species interactions to mediate ecosystem responses.

By robustness, we refer to a type of stability that focuses on the persistence of features of interest in a system in response to perturbations, particularly those not normally experienced by the system in its development or history. Thus, robustness is a useful way to think about ecosystem response to perturbations such as species loss. This section discusses a particular aspect of robustness: how differences in ecological network structure may mediate the degree to which primary species losses, however they are caused, may be followed by additional secondary extinctions. In a conservation setting, this type of research can help focus effective protection or intervention on less robust habitats, taxa, and ecological processes. This can help direct policy efforts and increased resources towards more fragile systems whose persistence may be relatively more threatened.

Assessing Ecosystem Robustness

Food webs provide a useful framework for beginning to assess the potential robustness of multispecies assemblages to biodiversity loss, as well as the impact of losing species with different functional or structural roles within ecological networks. One recent approach for assessing ecosystem robustness uses empirical data from ecological networks and makes use of simulation techniques from network structure analysis.

As shown in several well resolved food webs and pollination webs, secondary extinctions are much greater when species with many linkages to other species go extinct, compared to random loss of species. These studies suggest that regardless of the ecosystem, the order of species loss and the pattern of linkages among species strongly influence the rate of secondary extinctions. From a topological perspective, increased connectance across food webs buffers the impact of species losses: As connectance increases, the robustness of food webs also increases, because, on average, species have more potential prey items.

These types of analyses also need to be done with models that include dynamics among interacting taxa.

Results based on network structure may differ from what is seen in a dynamical framework. For example, a 20-species Lotka-Volterra dynamical food web has been used to study random versus ordered species losses in model systems. By assigning species a "tolerance" to a theoretical stressor and then removing species randomly versus based on their level of intolerance to the stressor, it was found that the average tolerance of remaining species is greater under ordered removals versus random removals.

Thus, robustness of the system to future change was greater under ordered loss of species. As with the previous network structure studies, the authors found that the order of species loss, given how species interact with each other as well as their tolerance to a perturbation, is very important for understanding secondary effects.

Similarly, the order of species loss was found to have a profound effect on the extinction dynamics of two large plant-pollinator communities. Risk is not equal for all species, but instead will be greater for species of high trophic position, rare species, and specialists. Pollinators are considered to be at higher risk of anthropogenic extinction than plants, due to their higher trophic position and other aspects of their biology.

Perhaps even more importantly, some authors also suggest that because the food web structure and dynamics change with each extinction, whether random or ordered, there is an unpredictability to compensatory dynamics that can change the potential impact of the future loss of a particular species. In their view, this unpredictability "argues for 'whole ecosystem' approaches to biodiversity conservation, as seemingly insignificant species may become important after other species go extinct."

This result is consonant with a finding from a previously described network structure analysis. Dunne and colleagues showed that while ordered removals of the least-connected species often lead to few secondary extinctions, in about a third of the webs examined such ordered removals resulted in relatively high secondary extinctions.

Unlike robustness of webs to loss of either most-connected or random species, which is positively correlated with connectance, robustness of food webs to loss of least-connected species is not correlated with any obvious network structure metric. Thus, the loss of apparently topologically insignificant species could lead to large negative effects, in a way not obviously predictable from the overall structure of the food webs.

Other models of food-web dynamics that go beyond equilibrium Lotka-Volterra-type dynamics to include more ecologically plausible assumptions about nonlinear, non-equilibrium dynamics, network structure and diversity, and include ecological and evolutionary adaptive dynamics will also be useful for exploring issues concerning the robustness of ecosystems to perturbations such as species loss due to habitat loss or climate change. The major challenge will be to bridge the gap between model and empirical ecosystems to usefully predict robustness of ecosystems to different types of perturbation and species loss.

Several interrelated research questions emerge out of the broad question of the robustness of multispecies assemblages:

— Do food webs from different habitat types (e.g., terrestrial vs. aquatic, desert vs. rain forest, fresh-water vs. estuarine vs. marine, boreal vs. equatorial, above ground vs. below ground) respond to perturbations in predictably different ways?

— How robust are food webs to different types of perturbation (e.g., habitat damage, loss and fragmentation; species loss/removal and invasion; chemical and climatic stresses)?

— How robust are food webs to varying perturbation intensity or order? To multiple perturbations? Are impacts likely to be synergistic? Are there thresholds of response?

— Ecological robustness can be assessed in many different ways (e.g., persistence of species diversity, retention of ecosystem function, low probability of destructive invasions, etc.). Do different types of robustness preclude each other in the form of trade-offs, or are they complementary?

— Ecological robustness can be assessed for different organisational levels and ways of grouping taxa (e.g., particular populations vs. a small set of interacting populations vs. a single trophic level vs. a food web for a particular habitat vs. a mosaic of dynamically shifting habitat types with loosely linked food webs). How do particular types of robustness, or the lack thereof, at different levels relate to each other?

— Are there general characteristics of the network structure, diversity, or dynamics of ecological networks that can be used to predict robustness to perturbations, or can be used to suggest ways to preserve or increase desired types of robustness?

Research that touches on some of these issues for diverse, multitrophic-level systems is just beginning, and there is much more work to be done. Most of the largely theoretical work concerning complexity-stability in food webs, and the experimental and observational work concerning diversity-ecosystem function helps identify the potential robustness of ecosystems to perturbations. However, both of these research approaches tend to focus on more narrow definitions of internal system stability rather than on system responses to external perturbations. Each approach also lacks the combination of empiricism and multitrophic-level inclusion that is necessary to facilitate advances that may help conserve natural systems and answer the robustness research questions posed previously. An example of how the two aspects can be fruitfully combined is research that used a theoretical bio-energetic food-web model, parametrised with data from a 29-taxa version of the Benguela marine food web, to look at the likely impact that culling fur seals would have on fisheries in that system.

Viable Areas of Food Webs

Ecology has contributed two basic insights into the biology of extinction: larger areas hold more species than smaller areas and larger populations persist longer than smaller ones. One key observation from field studies concerns the impact of the removal of top predators in unleashing top-down effects. The loss of such species can promote subsequent loss of further species and eventually lead to a so-called "ecological meltdown."

Such a phenomenon has been reported from predator-free forest fragments. The loss of predators generates strong imbalances shown by the disproportionate increase in the densities of prey and severe reductions of seedlings and saplings of canopy trees. An important issue here is that when looking at the space available for a given predator species, it is often assumed that it is the whole space that counts, while actually the space occupied by its prey is what counts. In this context, models reveal that the fragmentation thresholds experienced by a given species depending on a given resource are lower than one would expect by just considering the effective spatial area available. In other words, the patchiness denied by the spatial distribution of the resource strongly constrains the metapopulation dynamics of those species exploiting it and thus their effective fragmentation thresholds.

The relevance of topology in enhancing the effects of habitat loss and fragmentation can be observed by modelling different types of community structure. Melian and Bascompte have studied small food webs involving different types of architectures. They considered four trophic web modules: simple food chain, omnivory, apparent competition, and intraguild predation. Each module contained three trophic levels with three or four species. Again, it was shown that extinction thresholds are not only determined by life history traits, competitive colonisation abilities, and landscape properties, but also by the complexity of the food web. In particular, omnivory was shown to confer the higher persistence to the top species, while interactions between the two intermediate species decrease its patch occupancy.

Beyond the few-species model, other approaches have been taken using a number of community structures with different internal organisations, for example hierarchical communities of competitors at one extreme and neutral and quasi-neutral communities on the other. These have shown that community organisation is relevant in determining the effects of habitat loss and its spatial patterning.

Future research should address these topics by considering the food web structure and, in particular, should use approaches that seem to capture the fundamental traits of community organisation. One particularly difficult problem here deals with the possibility of considering all kinds of community interactions. How parasites and mutualistic interactions might modify previous predictions or alter the final outcome of ecosystem melting is largely unknown. Finally, when dealing with the sixth extinction at its largest scale, it is also important to provide useful tools for conservation efforts that deal with multiple scales and also include the intrinsic variability of climate and resources as an input.

Habitat Destruction and Food Web Theory

In the past few years there has been a revival of interest in the theoretical analysis of food webs, motivated in part by advances in the theory of networks. While a comprehensive food web theory or theories is still very much in development, progress has been made in developing models that describe the static, statistical properties of food webs. The most successful of these theories is the niche model, which assumes that predation occurs in an approximately hierarchical fashion.

Despite producing artificial webs in extraordinarily good accord with analysis of the largest available food webs, the niche model is not a panacea. The niche model does not purport to be a dynamic model, nor, despite the temptation, is it possible to equate niche value with a clearly measurable organismal trait such as body size. Further, there is no mechanism for maintenance of any particular food web pattern, either ecologically or over evolutionary time scales. All of these issues would be interesting courses for further theoretical investigation.

A standard approach to the study of food webs is the development of dynamical models describing changes in population densities in multitrophic systems. A similar and seemingly worthwhile variation would be to develop individual-based models in which the food web and its organisation constitute "emergent" phenomena resulting from competitive, predatory, and/or mutualistic spatially explicit interactions.

All of these models may be viewed, with some historical deference, as the next generation of food web models based on Lotka-Volterra-type interactions. The theoretical motivation for dynamic modeling is clear; food webs evolve on a variety of time scales, from the loss or introduction of a few species due to exogenous/endogenous causes, to wide-scale displacement of organisms because of climactic changes and/or anthropogenic disturbance.

Only a dynamic, spatially explicit model can hope to explain the underlying ecological reasons for the observed patterns. Indeed, habitat degradation is occurring at unprecedented rates; therefore, it seems irresponsible, even reckless, to wait the necessary time to agree on a comprehensive food web theory. While progress is made on general principles underlying a dynamic, spatially explicit, and possibly stochastic food web theory, modeling should attempt to offer practical recommendations and partial answers to the crucial questions in conservation and restoration ecology.

The most promising avenues for integrating reasonable notions of habitat degradation, an inherently spatial phenomenon, into food web models are outline below:

Non-random Destruction of Habitat

Complex ecological networks such as food webs arise from explicitly spatial and often heterogeneous interactions. The degradation of habitat is heterogeneous in its spatial distribution. Theoretical models of habitat destruction rarely incorporate such complexities, relying instead on simplified notions of disturbance amenable to computational and mathematical analysis. Some examples of theoretical approximations of disturbance include homogeneous site loss, homogeneous barriers, as well as spatially correlated degradation in models with exogenous and endogenous heterogeneity. Spatially realistic models of degradation in community/regional level food web models would be well served by borrowing from the field of metapopulation biology, in which habitat fragmentation is the norm and the methodology to deal with it has developed in kind. Examples of useful concepts include the use of incidence functions to describe the flux of immigrants between disconnected patches as well as the scaling of extinction probability with patch area.

The development of inexpensive and web-available remote sensing landscape data also bodes well for opportunities to integrate explicit land-use changes, habitat fragmentation, and vegetation changes into spatial models. This is relatively uncharted territory for theory and seems to require close affiliation between theory and empiricists working on local conservation and/or ecological studies. Finally, it is worth noting that not all habitat destruction is damaging. The pitch pines of the Pine Barrens are infamous for relying on forest fires to gain an advantage over other species that have diminished ability to re-sprout and recover after fires.

Taking a broader perspective, there is also evidence that at intermediate levels of disturbance the benefits of spatially limited refuges outweigh the costs to the community as a whole. In a sense, fragmentation can lead to a spectrum of solutions to the colonisation/competition trade-off, but in a completely connected world the best competitor wins. Deciding what degree of degradation a food web can sustain without collapsing is an important problem that theory is only beginning to confront.

Species-area and Link-area Laws

Food webs are typically denied at a .fixed scale, but the composition of local communities is determined by local as well as regional processes taking place at multiple scales. There is a widely held, though controversial, belief that local interactions and regional co-occurrence are responsible for the purported power-law scaling of species richness S with area 4, $S = cA^z$.

A recent theory proposes a means to extend the species-area law to a link-area law, that is, a higher-order indicator of the organisation of communities across scales. This theory generates many new questions for research. For example, to what extent will such a link-area law hold in a fragmented habitat? Available empirical data might permit computational removal experiments so as to examine the robustness of the link-area laws to artificial destruction of habitat. Also, is it possible to integrate the niche model into link-area laws, or at least to parametrise the niche model and test its predictions with this scaling theory?

Theory for Reserves

The flip side of developing theory for habitat degradation is the ongoing interest in the design of reserves. The basic scientific question in reserve design may be likened to that of a Lagrange optimisation problem from elementary calculus, wherein a student is asked to design a rectangular enclosure capable of holding the maximum number of cattle given a fixed amount of available fence. The answer is, of course, to construct a square enclosure.

Unlike this simple example, the tradeoffs in reserve design are myriad and far more complex and the "value" to be maximised is often contentious and under dispute. How many reserves should be set aside? Is it better for them to be contiguous or separated? What is the minimum useful reserve size and intra-reserve distance? How should the reserves be allocated dynamically in time?

Answering any/all of these questions is not strictly a matter of judging long-term sustainability of populations that depend on life-history, dispersal functions, evolutionary changes in organismal size, as well as on multispecies interactions. Benefits to tourism, public use constraints, and industrial access must also be considered. Some reserve design models incorporate explicit population dynamics, others do not.

However, marine food webs are rarely considered. Doing so would certainly add complexity to an already difficult problem. But in the long run, understanding the future of a single population requires at least some understanding of its interdependencies within the larger web.

Importance of Data on the Structure of Species Interactions

Data about the structure of species interactions will be invaluable in understanding key unanswered questions about how these interactions either respond to or mediate the effects of human perturbations. Even coarse data about known species interactions that are relatively easy to identify.

Biodiversity loss is a practical issue for conservation biologists and a problem for which solutions are needed in a short time frame. If food webs are to provide part of the solution, conservation biologists need a practical means of collecting food web data. Currently, putting together a food web is a time consuming task. Insects form the vast bulk of biodiversity and most of the recent high quality and highly resolved webs have been multiyear studies. The time and resources needed for this approach are simply not available to conservation biologists working at the coalface. Rapid assessment programmes were developed to provide a rapid, qualitative census of the biological resources of a natural area. While these represent a departure from the painstaking methods of food web construction, similar rapid techniques may need to be developed for conservation biologists who need species interaction data and who do not have the time/resources for a conventional approach.

A 'quick and dirty' approach to characterising interaction webs contrasts with the traditional structural food web approach of gathering detailed interaction data from very few sites. The more intensive and less extensive approach to compiling typical food web data is partly the result of the traditional focus by structural food web ecologists on identifying "universal" patterns of link structure across diverse ecosystems, rather than on applications for conservation. Early food web theory was criticised as being based on incomplete webs and this critique led to a more exclusive priority on gathering higher quality, detailed webs that could be used to test general theory about food web structure.

Useful ecological networks are relatively easy to construct, particularly by resource agency specialists who have an intimate knowledge of the natural history of their system, provided two conditions:

— They are explicitly acknowledged as works in progress, and the relative certainty of each link is clearly specified; and

— They are used primarily for comparisons within one system over space, time, or other conditions.

The first condition stresses the heuristic value of the process by considering each web to be an operating hypothesis that helps target further empirical or theoretical investigations. The latter condition facilitates comparisons by minimising variation among webs due to link uncertainties or investigator biases such as inconsistent levels of resolution within a web.

New ecoinformatic technologies for mining, managing, and analysing data will greatly facilitate the development of extensive ecological network databases that are relevant to conservation. Many conservation sites or large-scale monitoring projects have, or will have, extensive species lists stored in spatial mapping databases. Examples include: Sierra Nevada Forest Plan for bioregional monitoring and Millennium ecosystem assessment. For each species, attributes could be added that include its potential prey and predators from among the master species list, as well as species with which it is known to strongly interact in other ways. Then for any specified location, a potential network of interactions could be derived from this "meta-matrix" of interactions using the spatial information about which species coexist, and potentially interact, in that area. Thus, any interaction webs derived from this approach would be based on information compiled for all species found in the larger study area rather than on observations of interactions at each particular location.

While these literature-based data may be incomplete or problematic to use for testing general food web theory, the consistency in how link structure is specified across all samples would facilitate comparisons among sites and across gradients of habitat fragmentation. Selective detailed studies to quantify links at individual sites could test the validity of this "meta-matrix" approach.

Any developing species interaction database should include standards for specifying levels of certainty and sources for each link, and it should be flexible to accommodate new information that becomes available. Similarly, new developments in "semantic web" technology could facilitate data mining of dispersed web-based documents to compile food webs, and provide for data inference based on information from related species and systems to fill in trophic links where local data are lacking.

Ecological Network Model for Biodiversity Conservation

Ideally to restore ecosystems, habitats and species we would be able to protect vast areas from human exploitation. However the demands of economic development are such that this is not possible even in the last remaining areas of wilderness. The ecological network concept offers a way of reconciling these two conflicting demands by integrating biodiversity conservation with the exploitation of natural resources. This is done through embedding areas that are necessary to ensure the continued functioning of ecological processes or the viability of species populations within a wider landscape of semi-natural and managed lands.

Ecological networks contain four main elements:

— *Core areas*: These are areas where the primary function is biodiversity conservation. They are usually legally protected under legislation. These areas should provide a substantial representation of key natural or semi-natural ecosystems and contain viable populations of important or threatened species. Land use within these areas is managed to give priority to biodiversity conservation.

— *Corridors*: These are areas of suitable habitat that provide functional linkages link between core areas. For example, they may stimulate or allow species migration between areas. Corridors can be continuous strips of land or 'stepping stones' that are patches of suitable habitat. Using corridors to improve ecological coherence is one of the most important tools in combating the fragmentation that is threatening so many of Europe's habitats. Generally speaking corridors can be associated with higher levels of land use, as long as their function is maintained.

— *Buffer zones*: Protected areas should not be considered as islands that are safe from negative external effects. The resource use that occurs outside them can have serious impacts on species and habitats within, for example air/water pollution from industrial activities around a protected area can have serious effects on species inside it. Buffer zones allow a smoother transition between core areas and surrounding land use. The size and utilisation of buffer zones depends heavily on the particular needs of the specific ecosystem and its local population.

— *Sustainable use areas*: These are remaining areas that can come under more intensive land use. But they should still take full account of the successful provision of ecosystem goods and services.

The conservation of biodiversity has in recent years been confronted with two crucial challenges: first, the need to develop environmental management approaches that are effective in conserving biodiversity and, second, the need to accommodate the exploitation of natural resources where this is necessary to meet legitimate socioeconomic needs. Efforts to meet these challenges have encouraged the development of models that extend the scope of conservation actions beyond the traditional emphasis on the protection of threatened species and valuable sites. Significantly, despite the wide diversity of situations, means and perspectives in which conservation actions are being taken, two generic goals can be discerned in many of the newer approaches, namely a focus on maintaining the functioning of ecosystems as a means of facilitating the conservation of species and habitats and promoting the sustainable use of natural resources in order to reduce the impacts of human activities on biodiversity.

Various operational models have been devised to meet these generic goals. These models are known by a variety of appellations. Most common are ecological network, reserve network, bioregional planning and ecoregion-based conservation. To some extent these terms represent differences in scope or emphasis of the different models. However, all of these models aim to achieve the two generic goals noted above and can therefore be regarded as comparable approaches to the conservation of biodiversity. Here, 'ecological network' is generic term to refer to this broad category of conservation model. The use of a single term should not be understood as indicating a preference for a particular model, rather as a matter of convenience and also as a reflection of the terminology used by IUCN, which is giving special attention to this approach in its efforts to conserve biodiversity.

An ecological network can be regarded as acoherent system of natural and/or semi-natural landscape elements that is configured and managed with the objective of maintaining or restoring ecological functions as a means to conserve biodiversity while also providing appropriate opportunities for the sustainable use of natural resources. This definition has two key components: the maintenance of ecological functions as a means of conserving biodiversity and the sustainable use of natural resources. An ecological network aims to achieve these twin objectives by creating an infrastructure that facilitates ecological functioning but also accommodates a degree of human exploitation of the landscape where this is compatible with, or contributes to, the conservation of biodiversity.

The development of the ecological network model has its origins in the growing awareness amongst those actively involved in the conservation of biodiversity that:

- the protection of individual biological elements — predominantly a limited number of exceptionally valuable natural areas and threatened species—was not succeeding in arresting the decline in the integrity of the protected areas and many species populations.
- the viability of species populations is dependent on the existence of a particular complex of environmental elements and processes rather than on its simple isolation from human influences.
- the increasing extent and intensity of human activities in the landscape and their impact on biodiversity cannot be compensated through site protection measures alone.

Out of this awareness grew both a specific field of research—conservation biology —and various field projects which established the foundations on which the present array of initiatives are based. Prominent among these are five main groups of initiatives:

- UNESCO's 1974 Man and Biosphere Programme, which recognised the need to reconcile the conservation of valuable areas with local land-use needs through the delineation in Biosphere Reserves (currently over 350) of core areas, buffer areas and transition zones.
- flyway and corridor programmes that primarily aim to meet the needs of migratory species or those which need to disperse over large areas.

— ecological networks, which developed in Europe in the 1970s and 1980s where a strong land-use planning tradition had created the institutional environment for allocating functions at the landscape scale.

— reserve networks, which were developed in North America in the 1980s, primarily with the aim of conserving biodiversity at the regional scale.

— bioregional planning focuses on the process of planning and managing the protection of ecosystem services and biodiversity at the bioregional scale.

— ecoregion-based conservation aimed at conserving the world's key ecoregions, that is, relatively large units of land or water that harbour a characteristic set of species, communities, dynamics and environmental conditions.

At the strategic level, all these initiatives aim to ensure that ecological functions are maintained and natural resources are used sustainably. Athough the emphasis of the initiatives varies, a number of elements can be discerned which are common to all the approaches. These are:

— a focus on conserving biodiversity at the ecosystem, landscape or regional scale.

— an emphasis on maintaining or strengthening ecological coherence, primarily through providing for ecological interconnectivity.

— ensuring that critical areas are buffered from the effects of potentially damaging external activities.

— restoring where appropriate degraded ecosystems.

— promoting complementarity between land uses and biodiversity conservation objectives, and particularly by exploiting the potential biodiversity value of associated semi-natural landscapes.

These approaches do not, therefore, stake a claim to the landscape in a way that fences off the entire countryside from all human activities. They aim, rather, to focus conservation action on those areas and on those species communities that harbour environmental values which are crucial to the maintenance of ecological functions and, in the long term, to human welfare and also to delineate human activities in such a way that they are both economically viable and ecologically sustainable.

Although the first manifestations of the ecological network model date from the 1970s, it is only during the past decade that the approach has attracted sufficient attention and credibility to become widely applied. Today, well over a hundred examples of proposals, plans or projects for ecological networks are known to be under development or in course of implementation. These initiatives can be found on all continents, in a wide range of biogeographic zones and in landscapes varying from highly exploited to virtually pristine.

Indeed, although originally conceived as a means of reconciling the needs of biodiversity conservation with those of economic development in man-exploited landscapes in the developed countries, recent years have seen increased attention for the potential value of the model in the developing countries where there is an urgent need to secure sustainable development in regions characterised by a rich biodiversity and economically valuable natural resources, but where local populations are dependent for their welfare on the continued functioning of ecosystems.

14

Role of Women in Biodiversity Management

Women can play an important role in the biodiversity conservation and management. But, only in recent years has gender come to be acknowledged as an important variable in this sector. In community conservation efforts, there are clearly defined gender roles, particularly in the areas of plant and seed selection and preservation.

In a temple at Along in Arunachal Pradesh, for example, credit is given to a woman for domesticating rice. Women's role in biodiversity conservation has been overlooked, despite the fact that women have a profound knowledge of plants and animals in their environment. Although the natural resources of the environment provide the basis for both women's and men's livelihoods, women have traditionally used a variety of indigenous plants, trees and animals and so have a direct stake in their preservation.

Loss of habitats and biodiversity ultimately affects the underprivileged, the majority of whom are women. This is why women have participated in large numbers in movements like the Chipko Andolan. Although they possess knowledge about biodiversity conservation, poor women are often left with no choice but to exploit natural resources in order to survive. There is therefore urgent need for studies on the gender dimension in biodiversity conservation and management. Women, particularly in hill areas, have greater responsibility for agriculture than ever before, resulting from a process termed the feminisation of agriculture, in which women are increasingly responsible for agriculture.

The contributions of women and men to agricultural production are often divided along gender lines, with important implications for sustainable agricultural practices and biodiversity conservation. Men are generally responsible for land preparation, such as clearing and soil tilling, while women are responsible for sowing, hoeing, crop maintenance, harvesting, food processing, storage and seed selection for future planting. Studies show that because of women's responsibilities for securing food, fuel and water, they tend to have a greater interest in preserving and conserving croplands, forests and other natural resources.

Women are traditional caretakers of genetic and species diversity in agriculture; men are more often concerned with converting these resources into cash. At a meeting to consider methods of incorporating gender-sensitive approaches into plant genetic resource conservation and use, convened by the International Plant Genetic Resources Institute (IPGRI) and FAO in Rome in October 1996, the following working definition of gender was adopted: "Gender refers not to women alone but to the different socially constructed roles of, and relationships between, men and women across time and space as well as among variables of age, wealth, caste, ethnicity, etc."

The participants also concluded: "Within local communities, women maintain vital knowledge and play crucial roles in the conservation and use of biodiversity." An understanding of the different roles and responsibilities of men and women farmers in plant genetic resource conservation and management and the intrinsic value of their knowledge are crucial to initiatives on sustainable, effective and equitable plant genetic resource conservation and use. Agricultural research and extension based on local and gender-differentiated knowledge systems, institutions and management strategies should be provided to assist farming communities to maintain plant genetic resources in their agro-ecological systems.

Although the conventions on biodiversity and climate change and Agenda 21, approved at the June 1992 Earth Summit refer to the need for taking into consideration the gender dimension in environmental protection, the issue has not received sufficient attention at operational level. The Global Convention on Biological Diversity (CBD), ratified by 172 countries, recognises the "vital role that women play in the conservation and sustainable use of biological diversity" and affirms the "need for the full participation of women at all levels of policy-making and implementation for biological diversity conservation".

Situation before Implementation of the Project

The situation before implementation of the project can be described as having three levels, the third of which was addressed in particular by the project:

1. institutional arrangements for biodiversity management;
2. the role of the community in biodiversity management;
3. information gaps on gender roles and biodiversity management.

Institutional Arrangements for Biodiversity Management

India has one of the largest networks of protected areas in the world. Besides its national parks and sanctuaries, eight biodiversity-rich areas have been designated as biosphere reserves to conserve representative ecosystems. Five of the protected areas have been designated world heritage sites under the United Nations Educational, Scientific and Cultural Organisation (UNESCO) World Heritage List. Six wetlands have been designated under the Convention on Wetlands of International Importance, Especially as Waterfowl Habitat (Ramsar Convention) and a further 11 wetlands have been identified for intensive conservation and management.

Before the arrival of colonial rule, forests were managed by tribal and local communities. Wood and non-wood forest products were generally used in a sustainable manner, since people's lifestyles were simple and their needs were few. From the middle of the nineteenth century, however, the management of forests became the responsibility of government forest departments. Today, India has a well-developed institutional infrastructure for ex situ preservation of plants, forest trees, farm animals and fish. The Indian Council of Agricultural Research (ICAR) has established national bureaux of plant, animal and fish genetic resources.

The National Bureau of Plant Genetic Resources (NBPGR) has created, with assistance from the United States government, one of the world's largest repositories for the conservation of seeds of economic plants. NBPGR is a national organisation with a broad mandate to provide vital support in the form of required germplasm to the country's crop improvement programmes and to act as custodian of India's genetic resources of cultivated plants and their wild relatives. Besides its network of 12 regional stations and base centres in different agro-ecological regions of the country, it has strong links with:

— Crop-based ICAR institutes;

— National research centres;

— All-India Coordinated Crop Improvement projects;

— State agricultural universities;

— Relevant government departments;

— NGOs; research foundations;

— Private-sector research and development programmes.

The Ministry of Environment and Forests conserves forest genetic resources and the International Crops Research Institute for the Semi-Arid Tropics (ICRISAT) maintains a large *ex situ* collection of crop germplasm of chickpea, pigeon pea, sorghum, millet and groundnut. ICRISAT contributes to sustainable improvement in agricultural productivity in the semi-arid tropics through international research and related activities. It does this in partnership with national research systems and countries where its mandate crops are relevant to improving nutrition and wellbeing, especially among poor people.

The MSSRF selected specific locations in Tamil Nadu, Andhra Pradesh, Orissa and Kerala for intensive study under its Community Biodiversity Programme. Through this programme, scientists collected indigenous germplasm available within the study areas. In addition, interested NGOs also began sending valuable seed collections from their localities. During 1997-98, the number of entries in the Community Gene Bank increased to 679 and included cereal, millet, pulse and endangered and medicinal plant species. The entry register, herbarium, photographs and other essential information are regularly updated.

Community in Biodiversity Management

There are over 53 million tribal people in India belonging to 550 communities. Two-thirds of the tribal population live in central India, in the states of Madhya Pradesh, Orissa, Bihar, Gujarat and Rajasthan. Nearly 80 percent of the population of the northeastern states of Meghalaya, Nagaland, Mizoram and Arunachal Pradesh belong to tribal communities. In the northeastern tribal area, there are genetic resources of coir, jute, saccharum, cucurbits, rice, soybean, maize, citrus, buckwheat and several beans.

The central Indian tribal belt has contributed rich genetic diversity in minor millet, pigeon pea, rice, niger, sesame and forage grasses. In the Western Himalayan tribal belt, there is considerable genetic variability in buck wheat, amaranth, soybean, lentil, cowpea and pome and stone fruits. In the mountains of the southern Western Ghats, of which Wayanad, Kerala, forms an important biodiversity area, tribal people survive largely on subsistence agriculture. Concepts of purity and pollution govern women's roles and restrict entry to sacred groves.

Women play a critical role in conserving traditional varieties of food crops and medicinal plants. The varied landscapes of the Kolli Hills of Tamil Nadu are an area of considerable agro-biodiversity. Women here play a major role in seed selection and seed storage. Traditional systems of resource use and cultivation are now being replaced by cash crops in order to meet the people's needs. The forested hills of the Jeypore tract of Orissa are known for diversification of rice and home to different tribal communities. Both women and men are involved in conservation but their knowledge is being eroded.

Polygamy is widespread and some roles are demarcated by gender, although many activities are jointly undertaken in families. The Bhitarkanika sanctuary on the

Orissa coast is a protected delta area with considerable mangrove diversity. Caste and social divisions are of considerable significance in determining gender roles and relations. Sparsely populated Amnachal Pradesh has great ecological and socio-cultural diversity. The stratification, vulnerability and low status of tribal women could have a far-reaching impact on the prospects for biodiversity conservation. Sustainable use of natural resources needs further investigation.

In Mizoram, on the other hand, different topographical conditions and the practice of slash-and-burn agriculture have eroded biodiversity. New methods of organic cultivation, the outcome of a blend of traditional and modem practices, are being practised. The coral reefs and lagoons of the Lakshadweep Islands, together with the surrounding sea, provide a rich habitat for a unique matrilineal Muslim community. However, the status enjoyed by women because of the matrilineal system is circumscribed by the practice of women's seclusion, purdah and restricted physical mobility. Gender roles and the knowledge base of men and women are thus sharply defined.

Information Gaps on Gender Roles and Biodiversity Management

There is a widespread awareness of women's roles and gender-differentiated responsibility in managing biodiversity but there has been no systematic study to document situation-specific gender roles in biodiversity management, nor is there a database on the women professionals who contribute to biodiversity management.

Biodiversity in India

India is one of the world's 12 megadiversity areas, with over 45 000 wild plant species and 77 000 wild animal species registered, accounting for about 6.5 percent of the world's known wildlife. The sheer diversity and uniqueness are impressive. The biological wealth of India is a valuable asset to be managed prudently, particularly since the CBD recognises that the biodiversity present within the political frontiers of a country is the sovereign property of that country.

The diversity of living forms in India is the result of climate and soil variability and cultural and ethnic diversity. There are over 53 million tribal people in India, belonging to 550 communities. There are three megacentres of endemic biodiversity in India—the Western Ghats, the Eastern Himalayas and the Western Himalayas—and 25 micro-endemic centres. The total endemic taxa of flowering plants in the country number about 148 genera and 5 725 species, of which 3 471 taxa are in the Himalayas, 2 015 in peninsular India and 239 in the Andaman and Nicobar Islands.

According to MSSRF, available literature on gender and biodiversity provides the following indications:

— Men and women have different kinds of knowledge and information about plants and animals, in part because they have different tasks in farming and providing income and goods for their households.

— In many societies, women are mainly responsible for seed selection and storage, exchanging seed and ensuring that local agro-biodiversity is preserved.

— Both men and women preserve their native plant and animal species. The motivating factors, however, may differ. Men tend to be more interested in the market value of the species, while women may be more interested in their cooking and nutritional value. In areas where there is out-migration of men leading to the feminisation of agriculture, women tend to conserve a wide range of food and medicinal plants for ensuring household food and health security.

— The ways in which men and women have traditionally managed their seeds, plants and animals are fast changing. Government agricultural policies often lead to changes in dietary habits, resulting in tribal and rural families abandoning their traditional foods. This is particularly true in India, where the extensive public distribution system of the central and state governments generally concentrates on the supply of rice and wheat. Dietary changes also occur when tribal families lose access to non-wood forest products in habitats declared as protected areas.

— While women probably constitute the largest group of farmers who have conserved and improved ago-biodiversity, they often have no property rights to land. They also tend to be bypassed by development schemes related to biodiversity. Women's universities, home science colleges and other educational institutions catering to women's needs seldom include biodiversity as a field of study.

— The fast-growing literature on intellectual property rights, sui generis systems of plant variety protection and farmers' rights also reveals a lack of gender sensitivity in relation to the issues under discussion. There is practically no literature dealing with gender and intellectual property rights. A few studies that directly deal with gender and plant breeding, such as a number on wild food collection and land race selection, have a narrow focus with no broader interest in management systems. These studies often contain no comparison with the situation for men.

— Literature dealing with wild or domesticated plant genetic resources, when disaggregated by gender roles or the men-women division of labour, presents overwhelming evidence of a clear gender dimension to labour, knowledge and management. The same is true of agro-biodiversity in home gardens, forests and cultivated and fallow land. It is therefore surprising that the insights gained from over two decades of analysis of farm households and gender relations are not being applied critically to research on biodiversity. Most literature dealing directly with gender and biodiversity is based on platitudes and extrapolations rather than on empirical research or on empirical research that is superficial or narrow in scope. Broad generalisations are frequently made from scant evidence.

India's immense bio-geographic, socio-economic and cultural diversity cannot be described in any small selection of case studies. Each of the case studies referred to here covers a distinct ecosystem or region of significant biological diversity currently under threat. The five case studies carried out in Wayanad, Kolli Hills, Jeypore Tract, Bhitarkanika and the Lakshadweep Islands related to women and biodiversity conservation and sustainable use. The two case studies from Arunachal Pradesh and Mizoram in the north-eastern region were exploratory, included because the region is a significant biodiversity-rich area which today is in critical condition.

A survey of the literature was carried out as an initial step for the project. A brief communication was sent to women and men working in biodiversity at all levels throughout India, asking them to provide useful information for this project, including their curricula vitae indicating their area of specialisation, publications, years of work experience and so on.

The MSSRF project showed that gender roles in biodiversity management are socially constructed rather than biologically determined. The project found wide variations in the tasks assigned to men and women in different locations, with the same task, such as seed selection or winnowing, being done by women in some communities but not in others. In areas of traditional agriculture, among communities and classes that did not practise gender seclusion, women's participation in biomass-related activities was high and their knowledge and interest in conservation was apparent.

Women's involvement with conservation practices such as preservation of high-quality seed was high in communities where they were the main food producers. Examples are the Apatanis of Arunachal Pradesh and the Garhwalis of the Western Himalayas. The same is true of areas where women shared joint responsibility, such as among the Mizos, Nagas and some hill tribes of the Western Ghats. The project found that changes relating to age and gender were eroding this traditional knowledge base, for example in parts of Orissa. In agriculturally developed areas where market forces had penetrated deeply, such as Tamil Nadu, women were less involved in conservation practices but continued to have a role in seed preservation.

One factor that emerged clearly from the MSSRF case studies was that outside interest in local diversity is a significant element in motivating the interest of young people, women and men, in their own environment. This was particularly evident in the revival of local interest in agro-diversity in the Kolli Hills, in the mangrove diversity of Bhitarkanika and in the activities of the Young Mizo Association.

MSSRF case studies in Kerala, Tamil Nadu, Orissa, Arunachal Pradesh, Mizoram and the Lakshadweep

Islands, where distinct ecosystems and biological diversity are currently under threat, summarise the gender dimension in biodiversity management. The examples cited are based on innovative research carried out in ecologically rich and biologically diverse areas of India. The following sections highlight gender-based divisions of labour, knowledge, access to resources and roles in biodiversity management.

Kerala

In the Wayanad district of Kerala, men prepare the land and select the paddy seed. Older women may occasionally help but women in general are not allowed to do so. Women mainly carry out weeding, protecting plants from insect damage, harvesting and winnowing. Men and women jointly carry out post-harvest activities. While men make bundles of harvested crops, women carry them to the threshing yard. Women work alongside men cleaning and drying the grains but men alone are involved in seed storage.

Young and middle-aged women are not particularly involved in paddy seed selection and storage but are responsible for selecting and storing vegetable seed and tending vegetable gardens. Vegetables are sown during the month of kunni (August-September), a period in which women eat very well and proceed with full stomachs to the fields for sowing. Women use paroth leaves to protect the vegetable garden from pest attacks. Since all seed material is considered sacred, menstruating women are not allowed near any kind of seed store. Despite the oppressive concepts of pollution and purity to which they are subject, however, women do have a say in choosing which varieties of paddy are to be cultivated and in which fields.

Women take care to grow medicinal plants on their household premises. Both women and men of the Kurichiya tribal community have maintained the habitat for several species of medicinal plants through the preservation of sacred groves and associated areas. Kurichiyas use different plant species to make winnowing pans, baskets, grain storage units, umbrellas, mats, bows and arrows. These activities are carried out by men, since they are considered too strenuous for women.

Kolli Hills

Women and men jointly carry out agricultural activities, including land preparation. Ploughing is generally men's work but women also use the plough and there is no taboo against this activity. Women are absolutely prohibited from entering the sacred Grove of the Seven Deities at Mettuvillaram. Women are responsible for seed drying, cleaning, selection and storage and providing the correct quantities of seeds for sowing. For a mixed crop field, the women mix different crop seed material in correct proportions. Men normally carry out broadcasting and sowing in the early morning, while women are busy with household work. A clear division of gender roles may exist merely because of the conflicting demands on women's time.

Transplanting and weeding are women's jobs but both women and men participate in harvesting and threshing. Women have developed simple selection and storage methods for all food grains. The red earth coat, according to women, acts as a pest repellent. In the Kolli Hills, women pointed out that all grain seeds are dried in the sun and they were confident enough to claim that seed viability is maintained if dried seeds are kept in air-tight containers. Women use traditional wisdom and knowledge acquired from their mothers, older siblings and elders, to preserve part of the harvested grains as seed material for future use, thus contributing to conservation and maintenance of crop genetic diversity.

Women are also partners in deciding which crops to plant during the year. During the field study, MSSRF scientists often noticed men consulting with women or women interrupting men. Women were more confident than men when giving information to the scientists. There is renewed interest among both women and men farmers to revive and strengthen existing crops. Harvesting different fruits on household premises is done by either women or men, sometimes jointly. Young boys of around 15 are involved in the marketing of fruits; elderly people may also be involved.

Women's contributions are significant in food production and processing, although a male member of the family may occasionally help in processing food grains. Women normally work longer hours than men and make a substantial contribution to agricultural and food production. Women are more familiar with plant species useful for health care, such as edible greens and tubers. Women and men collect tubers together. Women, children and old men are involved in gathering firewood and grazing cattle.

Jeypore Tract

In the Jeypore tract at Koraput, Orissa, women and men jointly select the panicles of paddy but women have a greater role in the selection of minor millets and oil seeds. Men plough the land and sow the paddy seed, whereas women do weeding, transplanting and milling. Harvesting, winnowing and storage are done jointly by women and men. Women and men are equally involved in collecting, processing and marketing forest products such as grass, bamboo and resin. Seed storage is mainly a job for women but the storage structures are made by men.

In a few areas, there are taboos against women entering sacred groves on the outskirts of villages, in the foothills and on the hilltops. The management of herbal gardens is carried out by both men and women. All vegetable crops are cultivated by women but are jointly maintained by both women and men. Women take care of the post-production aspects of agriculture. In Koraput, collecting water is the duty of women and girls but women and men gather firewood. Grazing cattle and looking after them is the work of both women and men.

Bhitarkanika

In Bhitarkanika, women and men have knowledge of local health traditions. In particular, the MSSRF study showed that women's economic roles are determined by caste rather than gender alone. Oriya Brahmin and Kshatriya women are not allowed to leave their homes, thus effectively restricting their economic roles to domestic tasks. Oriya women of other castes, such as potters and weavers, and migrant Bengali women have freedom of movement. A different picture emerges from Bhitarkanika, where all agricultural activities — ploughing and field preparation, raising nurseries, transplanting, harvesting, threshing and storage — are done exclusively by men.

Women's only agricultural task is to boil rice before it is dried in the sun. Women are not permitted to enter sacred groves. Women with children aged between seven and ten were observed collecting fire-wood and fishing. During their field visit, MSSRF scientists observed that young men below the age of 25 served as field guides and knew all the medicinal uses of various plants and mangrove associate species. The roles of women and men in this region are governed first by the caste system and secondly by their economic conditions.

Arunachal Pradesh

The situation is somewhat different in the Eastern Himalayas, in the lower Subansiri district of Arunachal Pradesh, where firewood gathering is traditionally the task of men, although this is now beginning to change. There are no sacred groves in Arunachal Pradesh, except for small forest patches controlled by Buddhist monks, such as at Tawang. Most young women and men want to work on government farms. Ecological knowledge and skills in a tribe like the Apatani are clearly gender-based.

Older women identify varieties of paddy suited to different soil conditions and also select pure seed. Men, youths and younger women do not have this skill. However, Apatani women seem less knowledgeable about medicinal plants or less willing to share information. Household acclivities are sharpiy gender-based: food production and preparation are the responsibility of women, whereas hunting and fishing are the responsibilities of men and boys. In Apatani villages, men traditionally fetch fuel and firewood from the forest but women do so when necessary.

In Nishing villages it is rare to see women chopping wood early in the morning, as men are usually responsible for this task. Generally, the workload on old women farmers is increasing. In Arunachal Pradesh, all crafts are strictly gender-based. There is no cohesive approach to conservation and sustainable use and development of the rich resource base. The institutional infrastructure is ineffective and the administrative culture weak. There are no effective NGOs.

Mizoram

In Mizoram, the northeastern hills of the Pakhuii mountains are protected by their topography and by the local belief that the mountain is guarded by a spirit that will harm anyone who enters. Burning, forest clearing, protection and harvesting are done by both women and men. Women are fully responsible for sowing, weeding and storing. Site selection is assigned by drawing lots among male family heads only. All other tasks involved in the slash-and-burn cycle are done jointly or by women alone. Seed storage within the household is the responsibility of women but seed selection may be done by women or men.

Harvesting is done jointly. Women and men collect firewood and water. Men construct houses but women are sometimes involved in traditional technologies such as black-smithing. Small-scale trade is completely controlled by women. This spirit of cooperative organisation needs to be channelled both to improve women's status and to conserve the biological diversity of Mizoram.

Lakshadweep Islands

On the basis of interviews with ten women and ten men, it became clear that women show interest in planting trees of economic and food value around their homes, whereas men show interest in planting trees that provide fencing for their homes and in introducing ornamental garden plants. Coconut plantation, harvesting and pest control are carried out by men, while women do the early management of the coconut nurseries and assist the men in post-harvest operations. Both women and men spend at least two hours a day harvesting, tending and processing activities and both are conscious of the use of medicinal plants.

There is a story that when a botanist visited Minicoy and needed help from the local population to compile a list of all the useful plants, a woman told him his task would not be simple because he would have to list all plants, since there are no useless plants. Economic roles are sharply demarcated and women's and men's ecological knowledge is also distinct. Women are more conversant with the resources around their homes and along the reef and shore, while men, who go further to fish and collect coconuts, are more knowledgeable about the land, lagoon and sea.

Bibliography

Dutfield, Graham. *Intellectual Property Rights, Trade and Biodiversity: Seeds and Plant Varieties.* Gland/Geneva and London: IUCN and Earthscan, 239 p. 2000.

Falkner, Robert. "Regulating Biotech Trade: the Cartagena Protocol on Biosafety". *International Affairs, RIIA, Special Biodiversity Issue* 76 (2): 299-313. 2000.

Footer, Mary E. "Intellectual Property and Agrobiodiversity: Towards Private Ownership of the Genetic Commons". *In Yearbook of International Environmental Law, Volume 10 1999,* edited by Jutta Brunnée and Ellen Hey, 48-82. Oxford and New York: Oxford University Press. 2000.

Guay, Louis. "The Science and Policy of Global Biodiversity Protection". In *Governing Global Biodiversity: The Evolution and Implementation of the Convention on Biological Diversity,* edited by Philippe G. Le Prestre, 207-233. Aldershot, Hampshire: Ashgate Publishing. 2002

Khor, Martin. *Intellectual Property, Biodiversity and Sustainable Development.* Penang, Malaysia: Third World Network,123 p. 2002.

Le Prestre, Philippe G., ed. "*Governing Global Biodiversity: The Evolution and Implementation of the Convention on Biological Diversity*". Aldershot, Hampshire: Ashgate Publishing. 429 p. 2002.

__________. "Studying the Effectiveness of the CBD." *In Governing Global Biodiversity: The Evolution and Implementation of the Convention on Biological Diversity,* edited by Philippe G. Le Prestre, 57-91. Aldershot, Hampshire: Ashgate Publishing. 2002.

Le Prestre, Philippe. "The Operation of the Convention Governance System". *In Governing Global Biodiversity: The Evolution and Implementation of the Convention on Biological Diversity,* edited by Philippe G. Le Prestre, 91-115. Aldershot, Hampshire UK: Ashgate. 2002.

Makane Moïse Mbengue and Urs P. Thomas. "The Precautionary Principle: Torn between Biodiversity, Environment-related Food Safety and the WTO". *International Journal of Global Environmental Issues* 5 (1/2) Special Issue on the Precautionary Principle: 36-53. 2005.

McGraw, Désirée M. "The Story of the Biodiversity Convention: From Negotiation to Implementation". *In Governing Global Biodiversity: The Evolution and Implementation of the Convention on Biological Diversity,* edited by Philippe G. Le Prestre, 7-39. Aldershot, Hampshire: Ashgate Publishing. 2002.

Mishra, Jai Prakash. "Biodiversity, Biotechnology and Intellectual Property Rights". *Journal of World Intellectual Property* 3 (2): 211-224. 2000.

Pythoud, François and Urs P. Thomas. "The Cartagena Protocol on Biosafety". *In Governing Global Biodiversity: The Evolution and Implementation of the Convention on Biological Diversity,* edited by Philippe G. Le Prestre, 39-57. Aldershot, Hampshire UK: Ashgate. 2002.

Revéret, Jean-Pierre and Alain Webster. "Economics and Biodiversity Management". *In Governing Global Biodiversity: The Evolution and Implementation of the Convention on Biological Diversity,* edited by Philippe G. Le Prestre, 233-247. Aldershot, Hampshire: Ashgate Publishing 2002.

Thomas, Urs P. "The CBD, the WTO, and the FAO: the Emergence of Phytogenetic Governance". *In Governing Global Biodiversity: The Evolution and Implementation of the Convention on Biological Diversity,* edited by Philippe G. Le Prestre, 177-207. Aldershot, Hampshire UK: Ashgate. 2002.

Tuxill, John. "The Biodiversity that People Made". *World Watch* 13 (3): 24-36. 2000.

Index